Amoeba Management

The Dynamic Management System for Rapid Market Response

Amoeba Management

The Dynamic Management System for Rapid Market Response

Kazuo Inamori

CRC Press is an imprint of the
Taylor & Francis Group, an **informa** business
A PRODUCTIVITY PRESS BOOK

Original Japanese book edition published by Nikkei Publishing Inc., Tokyo, Japan. Copyright ©2006 by Kazuno Inamori.

Selections from Inamori Kazuo no Jitsugaku: Keiei to Kaikei Copyright ©1999 by Kazuno Inamori.

This English edition is published by arrangement with Nikkei Publishing Inc., Tokyo c/o Tuttle-Mori Agency, Inc., Tokyo

CRC Press
Taylor & Francis Group
6000 Broken Sound Parkway NW, Suite 300
Boca Raton, FL 33487-2742

Version Date: 20120629

ISBN 13: 978-1-4665-0949-8 (hbk)

Library of Congress Cataloging-in-Publication Data

Inamori, Kazuo, 1932-
Amoeba management : the dynamic management system for rapid market response / Kazuo Inamori.
p. cm.
Includes bibliographical references and index.
ISBN 978-1-4665-0949-8 (hardcover : alk. paper)
1. Management. 2. Leadership. I. Title.

HD31.I515 2012
658--dc23 2012015768

Visit the Taylor & Francis Web site at
http://www.taylorandfrancis.com

and the CRC Press Web site at
http://www.crcpress.com

Contents

Preface xi

SECTION I Amoeba Management

Chapter 1 Each Employee Will Take an Active Role in Their Own Respective Unit 3

The Origins of Amoeba Management 3

A Company Started with Seven Colleagues 3

Establishing a Management Rationale 4

Grow an Organization by Dividing It into Small Groups 6

The Three Objectives of Amoeba Management 8

Establishing a Market-Oriented Divisional Accounting System 8

Current Figures Are Needed—Not Past Statistics 8

The Basis for Making Decisions: "Do What Is Right as a Human Being" 10

"Maximize Revenues and Minimize Expenses" 11

Creating Divisional Accounting Systems Based on Truths and Principles 12

Direct Transmission of Market Movements Followed by Immediate Responses 13

Developing Leaders with Managerial Awareness 14

Colleagues as Business Partners 14

Achieving "Management by All" 16

Dissolving Labor–Management Confrontations with the Extended-Family Principle for Management 16

Sharing Our Management Rationale and Company Information Enhances the Managerial Awareness of Employees 19

"Management by All" Gives All Employees a Sense of Achievement 21

Chapter 2 Business Philosophy Is Crucial for Management 23

Divide into Smaller Managerial Units 23

Three Requirements for Dividing an Organization into Amoebas 23

Continual Review of the Organization 26

Determining Inter-Amoeba Prices 27

A Fair Judgment 27

Leaders Must: Management Philosophy 29

Conflicts of Interest Destroy Morale and Profits throughout the Company 29

Leaders Must Be Impartial Umpires 31

"Do Not Lie," "Do Not Cheat," and "Be Honest" 32

Embodying the Philosophy in Management 34

Place Highly Capable Leaders 35

Performance-Based Pay System and Human Psychology 37

Turn All Operations into Businesses That Cannot Be Imitated 38

Chapter 3 How to Organize Amoebas 41

Subdivide the Organization and Clarify Each Function 41

Functional Needs Are a Prerequisite for Forming Departments 41

Each Member of the Organization Should Have a Sense of Mission 42

The Subdivision of the Organization Must Satisfy These Three Conditions 43

Form Amoebas Where Overall Operations Are Apparent to Leaders 45

Assign Inexperienced People as Leaders and Train Them 47

Organizational Subdivisions Expand Business 48

A Flexible Organization Adapts to Markets 49

Establish an Organization That Is Well Prepared for Action 49

The Leader Is the Manager of the Amoeba 51

Management Rational Is Crucial for Autonomous Organization 51
Business Systems Administration Supports Amoeba Management....53
Role 1: Creating the Necessary Infrastructure for the Proper Functioning of Amoeba Management....53
Role 2: Accurate, Timely Feedback of Management Information 55
Role 3: Sound Management of Company Assets....55

Chapter 4 The Hourly Efficiency System: Focus on the Work Floor.... 57
Enhance All Employees' Profitability Awareness: The Rationale behind Unit-Specific Accounting....57
Use "Maximize Revenues and Minimize Expenses" to Simply Grasp Operations....57
A Management Accounting Method Suitable for Work Floors....58
The Difference between the Standard Cost Accounting Method and the Amoeba Management System....59
The Amoeba Form Becomes Visible....61
Amoeba Power Based on the Bonds of Human Minds....62
Creativity Comes from the Hourly Efficiency Report....63
Profit Management by Amoebas....63
Both Sales and Manufacturing Divisions Are Profit Centers....67
Always Express Targets and Results in Monetary Terms 68
Timely Awareness of Unit Profitability....69
Increasing Awareness of Time Management....69
Unifying Operations Management with the Hourly Efficiency Report....70
Fundamental Management Rules Based on Kyocera Accounting Principles 71
The Principle of One-to-One Correspondence 72

The Principle of Double-Checking Work........................73
The Principle of Perfectionism........................74
The Principle of Muscular Management........................74
The Principle of Profitability Improvement........................75
The Principle of Cash-Basis Management........................76
The Principle of Transparent Management........................76
Key Points of Performance Control........................77
Hourly Efficiency Reports Should Accurately Reflect Activity Results According to the Functions of Amoeba Units........................78
Hourly Efficiency Reports Should Be Fair, Impartial, and Simple........................78
The Flow of Business Should Be Presented as Results and Balances........................79
The Concept of Revenue Linking with Market Price: Three Methods to Recognize the Revenue of Amoebas........................80
The Custom-Order Method........................81
Adopting the Sales Commission Method........................83
The Flow of Figures Shows Market Movements........................84
The Stock Sales Business Method........................85
No Transaction at Cost........................86
Sales Is Responsible for Inventory Management........................87
Minimize Sales Expenses........................87
Internal Sales and Purchases........................88
Sales Department Commissions........................90
Review Profitability of Each Item........................90
Market Dynamism Is Formed In-House........................91
The Concept of Expenses: Correctly Perceive Reality and Control Expenses in Detail........................92
Recognize Expenses at the Time of Purchase........................92
The Beneficiary Principle of Expenses........................97
Handling Labor Costs........................98
Subdivision of Expenses........................99
The Concept of Time Focus on Total Working Hours: Bring a Sense of Urgency and Speedy Action into the Workplace........................100

Chapter 5 Fire Up Your Employees 103

Generate Profit with Your Own Will: Practicing Profit Management 103

The Fiscal-Year Master Plan 103

Aligning Our Mental Vectors by Setting Goals 104

Monthly Profit Management 105

Monthly Planning Based on the Annual Plan 105

Overall Verification of Running Totals 106

Sharing Goals within Amoebas 106

All Members Need an Understanding of Day-to-Day Progress 107

Execute Plans with a Powerful Desire to See Their Completion 107

Management Philosophy That Supports Amoeba Management 108

Pricing Is Management 109

Synchronizing Pricing and Cost Reduction 110

A Leader's Sense of Mission Is Essential in Adapting to Market Changes 110

Project Our Abilities into the Future 111

Running Long-Lasting Operations 113

Achieving Strong Cooperation between Sales and Production 114

Always Be Creative in Your Work 115

Set Specific Goals 116

Strengthening Each Amoeba 117

Sustaining an Awareness of the Entire Company 118

Leaders Should Take Initiative Rather Than Entrust Everything to the Work Floor 119

Cultivating Leaders 120

The Ultimate System to Elevate Managerial Awareness 120

Meetings Are Opportunities for Educational Guidance 121

Set High Targets and Live Each Day with Full Effort 122

Shared Awareness of the Significance of Decision-Making Criteria 123

SECTION II Questions and Answers for Management

Q&A Case Studies from Seiwajyuku Seminars 127

Afterword 145

Index 147

Preface

Since the Lehman Shock of September 2008, which was said to be a once-in-a-century event, the world economy has remained in an unpredictable state. Meanwhile, global market competition has continued to intensify. For a corporation to develop amid a relentless series of crises, the entire organization must unite in building an organizational structure that is strong and able to fight. The strength of the corporate structure must be raised to another level.

When I became involved in company management, I felt doubts about the state of long-standing, orthodox business management. When top management alone tries to exercise total control over a complex organization such as a large corporation, vigilance tends to weaken toward the outer edges—on the work floor. Weakened vigilance breeds various maladies, known collectively as *large-company syndrome*. Inevitability, the end result is deterioration of corporate profitability.

In contrast, the Amoeba Management System I developed through personal experience in management entails dividing a large organization into small units, called *amoebas*, with self-supporting accounting. A leader is appointed in each amoeba, and the company is thus run via a form of joint management. This kind of management method enables monitoring of every aspect of the company, and thereby detailed management of the entire organization. Even companies in which profitability has stagnated can then be transformed into previously inconceivable high-profit enterprises.

In the Amoeba Management System, management in line with the company's management direction is entrusted to amoeba leaders. In effect, amoeba leaders are running small organizations. Although they obtain approval from their superiors, amoeba leaders are responsible for preparing and then executing their own management plans. The Amoeba Management System can therefore cultivate leaders brimming with managerial awareness, even if they have little experience.

Centered on the leader, amoeba members set their own goals and put maximum effort into achievement in their respective work positions. The outcome is implementation of *management by all*, by which all amoeba members can concentrate their strengths on accomplishing goals.

In short, the Amoeba Management System is a management method that divides an organization into small units. Each unit is run using a market-oriented divisional accounting system linked directly to the market. While cultivating leaders with managerial awareness, within the company the Amoeba Management System achieves management by all, with all employees actively involved in management.

I developed the Amoeba Management System as "management accounting for leading business to success." As Kyocera's business activities diversified and steadily globalized, the application of Amoeba Management allowed us to continue practicing high-profit management.

Similarly, when building DDI (now KDDI) from scratch, I set up a divisional management accounting system based on the Amoeba Management System. The business environment of the communications industry is subject to sharp changes. Within such an environment, the divisional management accounting system has served as the basis for appropriate management decisions by KDDI, and therefore as the driving force behind the company's development.

In February 2010, at the request of the Japanese government, I assumed the post of chairman of Japan Airlines. My task was to resurrect the company after it filed for bankruptcy. Japan Airlines had struggled for many years with a deficit structure. Reviving the company required implementing a change of awareness through a management philosophy shared by all employees. At the same time, we introduced a divisional accounting system based on the Amoeba Management System. As a result, Japan Airlines achieved a V-shaped recovery in the following fiscal year, ending March 2011. After many years of generating huge losses, Japan Airlines achieved record profit.

Additionally, some 450 companies have received consulting services on Amoeba Management from a Kyocera subsidiary. Many of these companies are subsequently experiencing considerable improvement of business performance. As demonstrated by these examples, earnest study and implementation of Amoeba Management will result in rapid improvement of the management culture of an enterprise.

This book contains the essence of Amoeba Management, a management method into which I have put many years of careful thought and innovation. It was written for leaders active in diverse organizations, and people involved in business management and accounting. I would derive great pleasure from knowing it has helped to invigorate your organizations.

I earnestly pray that application of the Amoeba Management System by corporations and organizations around the world will further their development and help the people who work in them to lead yet happier, more meaningful lives.

Kazuo Inamori
January 2012

Section I

Amoeba Management

1

Each Employee Will Take an Active Role in Their Own Respective Unit

THE ORIGINS OF AMOEBA MANAGEMENT

A Company Started with Seven Colleagues

To help you understand Amoeba Management, please let me give a brief description of Kyocera's founding and its management rationale.

After graduating from the Department of Engineering at Kagoshima University, I joined Shofu Industries, an insulator manufacturer in Kyoto. In Shofu, I researched, and had succeeded in commercializing, a new field of ceramics. I had a difference of opinion concerning new product development with my boss, the newly appointed head of research. Unable to realize my goals as an engineer, I left the company.

Fortunately, I was supported to establish a new company. With seven colleagues who left Shofu Industries with me, I set up a company called Kyoto Ceramics. I did not have sufficient capital to start a company. Kyoto Ceramics was thus founded with financial support from many people who wanted the company to become a venue for presenting my technology to the world.

Had my family been affluent, or had I been able to create the company using personal funds, then the nature of the company would probably have been very different. However, I had no money, no experience, no advanced technology, and no sufficient facilities. What I did have was a company founded on the basis of a partnership with colleagues with whom I enjoyed mutual trust.

In starting the company, I became deeply indebted to Mr. Ichie Nishieda, who at the time was an executive director for Miyagi Electric Company. He said to me, "I invested money because I believe your thinking is sound and you have strong potential. When you start this company, you must not let money lead your management by the nose. You are to consider your *technology* as capital, and hold shares in your own name." Therefore, I started as a so-called owner-operator, holding "shares" of my investment by means of my ideas. The company began with this kind of wholehearted support, and the heart-to-heart bonds of colleagues sharing mutual trust became the cornerstone of Kyoto Ceramics' (and, later, Kyocera's) operations.

Being a novice in management, I was constantly worrying about "what the foundation for management should be." Before long, I began thinking about the importance of the human mind as a solid footing for management. This concept became Kyocera's foundation. The human mind is exceedingly changeable. Yet, once firmly bound, there is no stronger force in the world. A glance into history reveals countless examples of the great achievements made possible by the human mind. Ultimately, I concluded there is no surer way to lead a group than by reliance on the bonds of human minds.

Amoeba Management is based on the human mind. The human body has tens of trillions of individual cells, all of which work in harmony according to one will. Similarly, if thousands of amoebas in a company can unite their minds, then the company can function as one body. Even if there is occasional competition among amoebas, our mutual respect and support will allow the company to wield strength as a unified entity. Therefore, the bond of trust is an essential requirement for every person, from top management to members of each individual amoeba.

Establishing a Management Rationale

In the second year of the company, I hired about ten new high school graduates. A year later, just as they seemed to be getting used to the work, they suddenly came to me and demanded better conditions. They handed me their demands, which had been signed in blood.

Among their demands were "guaranteed wage increases and regular bonuses in the future." When I hired them, I told them, "At this point, I don't know how much we will be able to achieve. However, I intend to do my utmost to turn this into a fine company. Do you have the will to work in such a company?" Yet, after working for only one year, they threatened to quit if I did not provide guarantees for their futures.

I told them outright that I could not accept their demands. We had been operating for only about two years, and I still lacked confidence regarding the business. Giving guarantees for their employment in order to prevent them from leaving would have been deceitful. I told my young employees that I would give all my energy to provide them with future conditions even better than their demands.

Our discussions reached no conclusion at the company and continued late into the night at my home. They were stubborn and refused to yield. Over the following days, I tried to make them understand that I had not the slightest thought of personal gain as a businessman, but wanted to create a company in which each employee would be truly pleased to be a member. However, they were young and high-spirited, retorting that "capitalists and managers always put such a pleasant face on things to deceive workers."

At the time, I was regularly sending money home from my meager paycheck to help support my parents. I was the second son in a family with seven sons and daughters. After the war, we struggled in poverty. My brother and sisters had given up on higher education so that I could go to university. Unable to provide even for my own family, how could I guarantee the future of employees who happened to join my company? I felt an increasing sense that I had been allotted more than my fair share of obligation.

Even so, the company had been established. My benefactor, Mr. Nishieda, supported the company to the extent of putting up his estate as security. There was no way I could pull out now. Driven into a corner, I took up the struggle in earnest: "If you have the courage to quit this company," I said, "then why can you not find the courage to believe in me? I will stake my life to defend this company for you. If it turns out that I am running this company purely for my own self-interest, then you can put a knife through me and take my life."

Discussions continued over three days and nights, until, at last, everyone was convinced to stay. However, at the end of our negotiations, I could not help thinking anew about the significance of keeping a company in existence. Even now, as young employees joined, they were entrusting this tiny company with their lives.

For several weeks, I pondered the situation with a sense of gloom. Finally, I came to this conclusion: "I originally started this company to realize my dreams as an engineer. Employees now entrust their lives to it when they join. The company therefore has an even more important purpose. The purpose of the company is to safeguard employees and their families, and to aim for their happiness. My destiny is to take the lead and aim for the happiness of our employees." From that, I established the Kyocera management rationale as follows: "To provide opportunities for the material and intellectual growth of all our employees, and, through our joint efforts, contribute to the advancement of society and humankind."

The significance of Kyocera's existence as a company thus became clear: to provide opportunities for the material and intellectual growth of all our employees, and to contribute to society and humankind. Employees came to think of Kyocera as *their* company, and each person began working as earnestly as if he or she were the owner-operator. From about that time, my relationship with the employees was no longer one of management and labor. There emerged a genuine sense of fellowship among us, with each of us carrying the same degree of determination and sparing no effort in working toward the same goals.

Amoeba Management stimulates "management by all" due to its mechanism of self-supporting accounting for small groups. It is a management control system that concentrates the powers of all employees. However, this presumes the existence of a management rationale and philosophy that allow all employees to involve themselves fully in their work without doubt or suspicion.

Grow an Organization by Dividing It into Small Groups

Kyoto Ceramics soon began developing and commercializing many fine ceramic products that had never existed before—one after another. As a result, the company expanded rapidly. We started out with twenty-eight employees, and within five years we had more than a hundred; before long that number rose to two hundred and then to more than three hundred. Nevertheless, I was racing about overseeing all aspects of our operations, from product development to manufacturing to sales. Physically this was starting to take its toll, and the quality of my work itself began to suffer. It is often said, "A small company is like a balloon; if it grows too large, it will burst." In other words, when a small enterprise expands on the basis of rough estimates it will become unmanageable and finally collapse. Our company was already approaching that situation.

At the time, I saw that I might be able to solve the problem with knowledge of business management, organizational theory, and other topics useful for controlling an organization that had grown large. However, I had no basic knowledge of such matters, and furthermore, I was working late each night and had no extra time to study.

I was not even aware of the existence of the profession known as *management consulting*. Had I been aware, I might have scraped up the money from somewhere and sought guidance. In my ignorance, and with no resources to rely upon, I continued to worry alone about how to manage a swiftly growing company.

One day, an idea suddenly flashed in my mind: "While the company still had about a hundred employees, I could handle everything. What if I were to divide the company into smaller groups? At this stage we may have no leaders who can manage one hundred people, but we do have those who are becoming increasingly able to take charge of twenty or thirty people. What if I entrusted the management of small groups to people like these?"

By extension, if I were to divide the company into small groups anyway, why not adopt self-supporting accounting for each group? I could divide the company into the smallest possible groups capable of functioning as discrete business units, and place a leader at the head of each one. Each unit would then be administered independently, similar to small owner-operator workshops and factories.

Administering each organization using a self-supporting accounting system meant that we would need profit and loss statements for each one. However, specialized account statements were difficult for novices to maintain. I therefore adapted the profit and loss statements and devised the simpler *hourly efficiency report* for use by people with no accounting knowledge. As I will explain in detail (later in this chapter, and in Chapter 4), the report tabulates the implementation of the management principle "Maximize added value," which results from application of the basic principle "Maximize revenues and minimize expenses." The hourly efficiency report has a category equating to sales. Necessary expenses (excluding labor costs) are detailed underneath, and the difference is calculated. This mechanism allows us to recognize profitability at a glance.

The hourly efficiency report simplifies cost management for group leaders, enabling them to direct group members to reduce expenses in a specific area to increase the profitability of their department. Furthermore, the hourly efficiency report is clear enough for all group members to follow, opening the way for all employees to participate in management. In other words, while cultivating leaders, it raises interest in management and increases the number of employees who are interested in management and have managerial awareness.

Labor–management confrontation was still intense at that time—it was an era of frequent labor disputes. In such a climate, there was little leeway for thinking in terms beyond a confrontational structure between capitalists and workers. The common sense of the time was for managers to tell employees as little as possible about management, to prevent employees from attacking their weak points. Regardless of the times, Kyocera presented employees with transparent management in the form of the hourly efficiency report, and made the state of the company as open as possible.

Realizing that making the company situation visible to all could raise participation consciousness and stimulate ambition, I decided to place Amoeba Management at the heart of Kyocera's management control systems. After that, from the perspective of management control systems, Amoeba Management became the mainspring behind the rapid growth of our company.

The Three Objectives of Amoeba Management

Amoeba Management is not management know-how of the kind lionized by the business world. If it were simply management know-how, then it would be enough just to learn the methods and procedures. However, mimicking the methods of Amoeba Management gives little guarantee of a successful result. Why? Because Amoeba Management is a total management control system, based on a philosophy of management and intimately tied to all systems within the company. Since it is closely associated with all aspects of management, giving a clear perspective of Amoeba Management as a whole is difficult. Therefore, when learning about Amoeba Management, a good understanding of the objectives is essential.

In general terms, Amoeba Management has three objectives:

1. Establishing a market-oriented divisional accounting system
2. Developing leaders with managerial awareness
3. Achieving "management by all"

Now we will discuss these objectives one by one.

ESTABLISHING A MARKET-ORIENTED DIVISIONAL ACCOUNTING SYSTEM

Current Figures Are Needed—Not Past Statistics

Like most companies, Shofu Industries—the company I first worked for after graduating from university—had various administrative departments:

Accounting, General Affairs, Personnel, and others. Specialized work was entrusted to its relevant department. My job was to concentrate on research and development, manufacturing, and new-product sales. As for accounting, all accounts associated with my division were handled by the Accounting Department. I had no hand in it at all.

When I left Shofu at age twenty-seven and founded Kyoto Ceramics, I was a complete novice at management. Mr. Masaji Aoyama, my superior at Shofu and a benefactor in establishing my new company, gave me an overall introduction to accounting. Mr. Aoyama was general manager of the administrative division at Shofu and was well versed in cost accounting.

Several months had passed since my company had been founded. Mr. Aoyama performed our cost accounting while processing the day-to-day accounting slips. He showed me the results, intending to explain in detail the manufacturing cost of goods shipped three months earlier.

At the time, I was rushing about all day, solely responsible for all aspects of the business, including product development, manufacturing, and marketing. I had no time for a leisurely look at costs from several months before. While giving agreeable responses, I took little notice of what he was explaining. No doubt, Mr. Aoyama felt I was thinking lightly of cost accounting. Nevertheless, he came to me repeatedly to show and explain cost tables.

Somewhat perturbed by his persistence, I said, "Mr. Aoyama, these past figures are of no use to me. Knowing the cost months after the product has been shipped is of no value. I am working each day to produce profits just for this month. Even if I am told these are the figures for several months ago, there is nothing I can do about it now. More importantly, the market for electronic components is highly competitive. Prices for orders received today are falling. Historical costs have little significance."

Thanks to Mr. Aoyama taking on the work of accounting and administration, Kyocera started off well. Wanting me to be aware of the importance of cost accounting, he came repeatedly with figures and explanations. I truly regretted my impertinence, but I still could not rid myself of the notion that in totaling costs after the fact, he was merely presenting me with a record of my management several months before.

As the field of ceramics was still entirely new, monthly repeat orders were rare. Our company was accepting orders for and delivering brand-new products, and then accepting orders for the next new product. When repeat orders did arrive, fierce competition resulted in consistent demands for lower prices. Not unlike today's deflationary economic environment, market prices continued to fall, and cutting prices was a matter of course.

In many cases, by the time cost accounting figures arrived several months later, we were no longer manufacturing the product, making the figures essentially useless.

Most industrial products pass through many manufacturing processes on the way to completion. Raw material costs, labor and subcontractors, electricity, depreciation, and other expenses incurred are totaled as production costs. On the other hand, the selling price is set by the market, regardless of our cost. Our profit is the difference between the two. Moreover, the market price paid by customers is certainly not uniform. There is no guarantee that we can obtain orders this month at the same prices as last month. Especially in today's era of sharply falling prices, the selling price is effectively declining on a daily basis.

In these circumstances, conventional cost accounting is not useful. Basing management on cost data relevant to conditions from months earlier is not effective for dealing with constantly changing market prices. We need timely cost control in manufacturing for a rapidly changing market. The manager needs to know the *current* operational status of the company. We need "live figures" to make effective decisions.

The Basis for Making Decisions: "Do What Is Right as a Human Being"

Around the time I started the company, like it or not, I was called upon to make all kinds of decisions as a manager. This being a start-up venture, one wrong decision could push the company off the rails. Day after day, I continued to rack my brains over some kind of criteria to adopt for making decisions.

At the end of all the worry, I realized that what was good for day-to-day life should also be the basis for management decisions—in other words, "Do what is right as a human being." If we operate in violation of generally accepted ethics and morals, then we cannot expect to do well in the long term. Therefore, I felt the basis for decisions should be the concepts drummed into us when we were children: what is right and what is wrong for human beings.

"Do what is right as a human being" became our fundamental principle for management and the criterion for all decisions. Specifically, this embraces impartiality, fairness, justice, courage, good faith, endurance, effort, kindness, compassion, modesty, philanthropy, and other elements of generally accepted universal values.

Because I was ignorant of management techniques, I had no notion of the so-called common sense of management. All decisions had to be considered in terms of basic principles. However, this led to the recognition of crucial management principles.

"Maximize Revenues and Minimize Expenses"

Here is a good example. Not long after the company was founded, I had our accounts examined by an experienced accountant from Miyagi Electric Company, a man to whom I was indebted in many ways. I asked him how the accounts for this month were doing. He explained using difficult accounting terminology. Being ignorant of that field, I did not fully understand. After repeating the question again and again, I finally said, "I understand. In short, all we need to do is maximize revenues and minimize expenses. If we do that, profits will naturally increase."

Because I was still a novice at management, I probably saw the essence of things in a simplified form. I realized that "Maximize revenues and minimize expenses" is a fundamental principle. Since then, in keeping with this fundamental principle while single-mindedly striving to maximize sales, I have also endeavored to reduce expenses wherever possible. Consequently, as I described earlier, the business expanded rapidly and profits rose.

In hearing this, somebody will invariably say, "Of course. That's obvious!" Yet, this principle reaches beyond common sense and into the heart of management. In conventional enterprises, whether in manufacturing, distribution, or services, management is based on the tacitly accepted notion that "this industry has such-and-such a profit ratio." For a manufacturer, the profit ratio is several percent; for the distribution industry, 1 percent is considered good. Management is based on the common sense of the industry. If results match common sense, people will think they have "done well."

Yet, from the perspective of the principle "Maximize revenues and minimize expenses," the potential for sales is unlimited, and expenses can be cut to an absolute minimum. Profit can ultimately be raised to any level.

Furthermore, increasing revenue is not a simple matter of raising prices. As I will explain in Chapter 5, the key to increasing revenue is to start from the fundamental principle that "Pricing is management," and focus on finding the maximum price that customers will be happy to pay. Similarly, when reducing expenses, do not give up when you feel you have reached the limit. At that point, you need to believe in the infinite possibilities of humankind and apply even more unrestrained effort. Doing so gives the potential to raise profits to whatever level you desire. On the basis of this fundamental principle, with the persistent efforts of all employees, an enterprise can continue to produce high profits for a long time.

Creating Divisional Accounting Systems Based on Truths and Principles

Realizing this principle while managing the company, I put all my energy into turning Kyocera into a high-profit enterprise. However, before long, as the company grew I began to feel a certain anxiety. As the manager, I was able to run the entire company in line with the principle "Maximize revenues and minimize expenses." Yet, as the organization grew, I began to feel anxious about the limitations of my ability to apply this principle to every corner of the company.

As the all-important revenue and expenses are generated at different work floors on a daily basis, employees at each site need to understand and apply the principle. At the time, most employees worked in the Production Department. Although cost reduction was a viable proposition, there was little interest in or sense of responsibility for increasing revenue. Working from the principle, however, we must strive to both minimize expenses *and* maximize revenue simultaneously at every processing stage. Unless all production leaders were to link their work to increased revenue, the will to maximize revenues would be unlikely to emerge.

Moreover, I feared that as the organization grew too large, minimizing expenses would give way to roughly estimating expenses. Eventually it would become difficult to determine what expenses were being generated where. We needed a control system to allow detailed awareness of accounts. Subsequently, it occurred to me to divide the entire company into small unit operations, which could buy and sell amongst each other within the company.

For example, fine ceramics production can be divided into separate stages, including materials preparation, forming, sintering, and processing. If we separated each process into an operating unit that could trade with other units, the Raw Material Division would generate a "sale" when it sold to the Forming Division, whereas the Forming Division would record a "purchase." In other words, by adopting a system of buying and selling work-in-process between the actual production processes, each process becomes an independent accounting unit, just like a small enterprise. Each process is thus able to function independently, all the while realizing the fundamental principle "Maximize revenues and minimize expenses." At Kyocera, this is called Internal Sales and Purchases, and it is a primary feature of Amoeba Management.

Turning the company into a set of small operating units enables management to accurately assess where profits are made and where losses are generated, in relation to overall company profitability. This places top management in a much better position to make the right managerial decisions, and it facilitates the fine-tuning of company administration as a whole. The divisional accounting system for small groups can thus be regarded as the prototype for Kyocera's Amoeba Management System.

Direct Transmission of Market Movements Followed by Immediate Responses

To apply the principle "Maximize revenues and minimize expenses," I subdivided the organization and turned each area into an independent accounting unit—an amoeba. A leader is put in charge of each amoeba and entrusted to manage it. Although final approval by superiors is necessary, overall management is delegated to the leader. This includes business planning, performance control, labor management, and purchasing.

Although the amoeba is a small organization, management nonetheless requires involvement in accounts and a minimum level of accounting knowledge. However, at Kyocera in those days, not all amoeba leaders had the necessary knowledge. Therefore, we had to develop a simplified accounting mechanism for amoebas, one that could be followed without specialized knowledge. This is where I came up with the hourly efficiency report (detailed in Chapter 4).

In addition to detailing revenue and expenses for each amoeba, the hourly efficiency report measures the difference between the two. The difference, which equals the added value, is divided by total labor hours to obtain added value per hour. This system makes it easy to see just how much hourly added value an amoeba has produced. Moreover, comparison of the projected and actual results of the hourly efficiency report gives amoeba leaders a timely grasp of the state of progress, through comparison with a previously prepared sales plan, production plan, expense plan, and so forth, allowing them to take whatever steps may be necessary without delay.

Market prices change hour by hour. Without a flexible response to these changes and constant efforts to stay ahead, it is not possible to achieve our added-value and profit targets. For this reason, I divided complex production processes into many small amoebas, while also developing a management control system that allows amoebas to trade with each other and gives real-time awareness of each amoeba's actual results.

Using such a management control system, a sharp decline in market prices is quickly reflected in the buying and selling prices between amoebas. The amoebas affected can then take immediate steps, such as reducing expenses. In other words, not only are market dynamics transmitted directly to every corner of the company, but also the company as a whole is able to implement a timely response to market changes.

Moreover, internal transactions have a substantial effect on quality control. The purchasing amoeba cannot buy internally unless quality requirements are met. Therefore, work-in-process that does not meet predetermined quality requirements at any particular stage will not move to the next process. Each internal transaction is thus a quality inspection gate. As a result, amoebas at all processing stages become keenly aware of the need for quality assurance.

To develop market sensitivity and an ability to respond to these rapidly changing conditions, an organization must remain flexible. It must be free to divide, consolidate, or grow as needed for business development.

The name *amoeba*, as applied to Kyocera's operating units, came from a Kyocera employee who likened the flexibility of the company's small organizations to the repeated cell division of living amoebas. An amoeba as an accounting unit has clear-cut intentions and targets. It functions as an independent organization and aims for continued growth.

The truths and principles of company management are to maximize revenues and minimize expenses. Applying and practicing this principle throughout the company require us to divide the organization into small units. Each unit undertakes divisional profit management and is thus able to respond immediately to market movements. This is the first objective of Amoeba Management.

DEVELOPING LEADERS WITH MANAGERIAL AWARENESS

Colleagues as Business Partners

When Kyocera was founded, I was in direct control of all departments, including Development, Production, Sales, and Administration. If a problem occurred on the production floor, I would rush over to give direction. I visited clients to obtain orders. When a complaint came in, I would take charge of the response. At the time, I had to fill all of these functions alone. Amidst this hectic lifestyle, I seriously thought about how much it would help if I could clone myself and order one of me to visit customers and take care of sales, another to solve production problems, and so forth.

Yet, being busy was not the sole problem. In every company, the top executive is a solitary being. As the top person in the group, he or she has the final say and must take final responsibility. There is always a sense of loneliness. In my case, due to not having any experience in running a company, this feeling was that much stronger. Deep within, I wanted partners with whom I could share the feeling of responsibility, as well as the exhilaration and anguish of running a company.

While an enterprise is still small, the top executive, even if extremely busy, is capable of overseeing the entire company. However, as the company expands, it becomes increasingly difficult for one person to maintain oversight of production and sales, development, and all other functions. At that point, the general practice might be for the top executive to divide the organization into a production department and a sales department, giving overall responsibility for one department to another person.

Even so, if the business continues to expand, then it gradually becomes difficult for one person to oversee an entire department, be it Sales, Production, or another function. In these circumstances, Sales, for example, might be divided into regional groups—East Japan Sales and West Japan Sales. As the client base grows, West Japan Sales might be further divided into subregions—such as Kansai District, Chugoku District, Shikoku District, and Kyushu District. In Production also, if you wish to maintain detailed awareness of costs and profits, management by one person alone becomes impossible. The next logical step might be to subdivide Production into organizations for different product types, processes, and so forth.

The scale of the enterprise grows, and it becomes impossible for the top executive and leaders of individual departments to oversee the entire company. By further dividing the organization into smaller operating units, with each unit running under self-supporting accounting, unit leaders can maintain an accurate awareness of the state of their respective units. Furthermore, since leaders of the operating units are in charge of small groups of personnel, they can easily maintain control over the progress of day-to-day work, their own processes, and other organizational operations. They are able to adequately manage their respective operations even without high-level administrative skills and specialized knowledge.

That is not all. Even in small operating units, entrusting management to leaders stimulates managerial awareness and leadership development. This creates in them a sense of responsibility as business leaders and spurs further effort to improve performance. The difference here is in perspective, which shifts from the "passive" mentality of an employee to the "active"—"I will do this for the benefit of my unit"—mentality of a leader. This change in perspective marks the beginning of managerial awareness.

In consequence, the attitude changes completely from that of the wage laborer who works for a specified time period to receive a specific amount of compensation, to that of one who works toward the compensation of the members of the group. At this point, the individual becomes willing to work to improve operations, even if it involves self-sacrifice. Ultimately, business partners who are ready to share the burden of management responsibility emerge one after another from among unit leaders.

Thanks to Amoeba Management, the Kyocera Group has produced many leaders who possess the consciousness of business partners. From the introduction of Amoeba Management to the present day, Kyocera amoeba leaders have demonstrated superb management.

Divide the organization as necessary into small units, and rebuild it as a unified body of small enterprises. Entrust the management of these units to amoeba leaders to cultivate personnel with managerial awareness. This is the second objective of Amoeba Management.

ACHIEVING "MANAGEMENT BY ALL"

Dissolving Labor–Management Confrontations with the Extended-Family Principle for Management

After World War II, Japan became a democratic nation, and the Communist Party, illegal before and during the war, was resurrected. There was also reaction to prewar suppression; and after the war, socialist strength increased sharply. Labor disputes were frequent.

Reformist influences were strong, particularly in Kyoto, where governors affiliated with the Communist Party led the prefecture for decades after the war. Workers insisted on their own rights only and showed little understanding of the worries and pains of managers. Meanwhile, there were some managers who retained prewar attitudes and regarded workers as little more than tools to be put to work. Since Kyoto was fortunate enough to escape U.S. bombing raids, the townscape and local people remained largely unscathed. This was perhaps a factor behind the retention of old attitudes toward workers.

When did such hostile labor relations emerge? An examination of labor relations in the context of human history brings up a number of thoughts. At the dawning of civilization, people began shifting away from the migratory lifestyles of hunters and gatherers to the settled lifestyles offered by agriculture. With the beginning of agriculture, people introduced more stability into their lifestyles by putting away food provisions for use in case of natural disaster. With that, an excess of stockpiled food could be taken and sold to people in areas experiencing food shortages. Merchants with abilities in business began to emerge.

The first merchant businesses were run by individuals and families. As their businesses grew, they began hiring workers who were not part of the family. For the first time, there arose a relationship between management and labor. Commerce prospered with the ages, and merchants with extensive dealings hired more and more workers. At the same time, there appeared merchants who endeavored to pay their workers as little as possible so as to maximize profits. As such merchants grew in number, commerce began to experience a conflict of interest between managers and workers.

In time, the concept of capitalism developed, and manufacturing and other diverse kinds of industry emerged, leading to the present-day system of publicly traded companies and other forms of corporate organization. All the while, business organizations grew in size. As operators became busier than ever, they might have divided a company's operations, appointing executives and partners to whom they could allot part of the work and responsibility.

It may be true that increasing the number of executives to share the CEO's management responsibilities improved efficiency. However, the executives and other managers usually numbered in the dozens at most, and were vastly outnumbered by the labor force. This did nothing to eliminate confrontation, which simply grew in severity.

Workers tend to push their own rights and do not try to understand the worries and pains of managers. Similarly, managers tend not to understand the position of workers or try to safeguard workers' lifestyles and rights. Because each side tries to protect its own interests and does not try to see from the perspective of the other, the conflict between labor and management gradually intensifies.

World War II ended, and not long afterward I established a company in Kyoto, an area that was experiencing growing confrontation between management and labor. I don't know whether it was because they had been raised in such a climate, but, as indicated by the confrontation that I discussed in the "Establishing a Management Rationale" subsection of this chapter, many of the workers who joined the company did not trust top management and managers, and were convinced that management was hostile to workers.

At the time Kyocera was a small start-up business. For the company to survive, it was essential for all workers to unite in order to cope with stiff market competition. At that time, if the enterprise were consumed by internal labor–management conflict, its very existence would be imperiled. By one means or another, we had to have a company culture free of internal conflict, where management and labor could work together as a unified body.

I spent a lot of time thinking about how to solve this problem. The conclusion I reached was "If the manager values the viewpoint and rights of the workers, and if workers possess the same attitude as the manager to contribute to the entire company, then the labor–management conflict should dissipate naturally."

Business enterprises exist in various forms—as private corporations, limited liability companies, publicly held companies, and so on. It occurred to me that if we could build a company in which "all employees are managers," then no labor–management confrontation could exist. Nothing could be stronger than an enterprise in which all employees united for the company's development.

At the time, I was aware of a type of business entity known as a *partnership* adopted by accounting and law firms in the United States. Partners in such arrangements ran their businesses with joint responsibility and liability. I considered the possibility of making all Kyocera employees partners in the company. Unfortunately, that type of business arrangement did not exist under Japanese law.

Even so, believing in an environment in which all workers shared the same goals as management and regarded mutual cooperation as the ideal, I then looked to the traditional Japanese family for a model. The word *family* is used here for the traditional family group in which grandparents, parents, and children all do their utmost for the good of the family. Parents are considerate of their children; children are considerate of their parents. The family is a community held together by destiny, with everyone obtaining a sense of enjoyment as the family grows and develops. Such a family relationship is one of mutual love, with members ready and willing to do anything for each other. This is what I mean by the *extended-family principle*.

If the company becomes a community held together by destiny, as one extended family, and if mutual understanding, encouragement, and help are freely given among managers and employees, just as in a family, then it should be possible to run the company with management and labor as one body united for the same purposes. Even in the face of severe market competition, operations should naturally proceed well, as united efforts are directed toward the development of the company. I named this concept the *extended-family principle* and made it part of the company's management foundation.

At a time when labor–management confrontation was regarded as inevitable in Japanese society, I aimed to build human relationships between managers and employees, similar to what one might find in a family. I aimed to build a workplace in which as many employees as possible took an active part in running the company.

Sharing Our Management Rationale and Company Information Enhances the Managerial Awareness of Employees

Even so, regardless of how strongly the extended family principle is advocated, that alone will not eliminate confrontation between labor and management, nor will it foster a climate of labor–management cooperation. To rise above differing labor–management viewpoints and unite all employees, it is essential to have clear management objectives and a management rationale that all employees can firmly support.

In many cases, family businesses are succeeded by children or founded by people who simply want to make money. If Kyocera had been such a company, it would surely have been very difficult to turn it into a company where labor and management are united. However, Kyocera was established by a group of like-minded people who shared mutual trust from the start. As the manager, I never had the slightest thought of making money for personal gain.

In addition, through the process outlined in this chapter, we officially made it our management rationale: "To provide opportunities for the material and intellectual growth of all our employees, and, through our joint effort, contribute to the advancement of society and humankind." Therefore, because the company exists for the purpose of providing opportunities for the material and intellectual growth of all employees, there was no conflict of interest in unifying labor and management to work for the development of the company. A universal management rationale that could be shared and accepted by all employees became fertile ground for a corporate culture in which labor and management rose above conflict and united for the same purposes.

Moreover, establishing this management rationale has made it possible for me, as a manager, to speak frankly to employees. Had I been in business merely to make money, my aim would likely have been to exploit the workers and make them work for my personal gain. In the case of Kyocera, as the manager, I have endeavored to give all my energy to leading the company for the happiness of all employees, regardless of my own personal sacrifice. Because I was trying to make the business succeed for the good of all, I was able to reprimand and scold where necessary without reservation. Even so, a feeling of kinship arose among employees because we were all working toward the same goals.

Despite all this, it did not follow that all employees understood my difficulties as a manager. I might have said to an employee, "It's not the time to be doing this. The company is hardly in the right situation for this kind of thing." A thorough understanding of what I meant did not emerge immediately. A psychological gap remained between the employees and me. It then occurred to me that the reason they did not understand what I was saying was that they were not aware of the realities of the company's circumstances. I thought that if I stepped right out and informed everyone of these realities, then they would perhaps have a better understanding of my situation as the manager. As mentioned, managers generally do not try to think from the viewpoint of the workers, and workers tend to insist only upon their rights. The confrontational structure persists. However, I wanted all employees to adopt a manager's mentality and level of awareness. This led me to conclude that the most important step was to disclose as much information as possible about the company's circumstances, to let everyone know about the business matters that were troubling me and reveal all problems without embarrassment.

At that time of intense conflict between management and labor, it was normal for managers to tell workers as little as possible about the company's financial circumstances. Amid these social conditions, I decided to inform all employees of the realities of the company, concealing nothing, and thereby gain their understanding. If everyone understood the realities and problem areas of the company, my worries would be shared with all employees. This would have the effect of cultivating a manager mentality in all employees.

"Management by All" Gives All Employees a Sense of Achievement

Under Amoeba Management, the company is divided into small units. Oriented around a unit leader, all members take part in managing the unit. In the process, key information about the amoeba's operating conditions and those of the company as a whole is disclosed in full to all employees during our morning meetings and through other means. The maximum possible disclosure of company information in this manner both facilitates full-participation management and prepares the ground for independent participation in management.

With active participation in management by all employees, and with each person prepared to independently fulfill his or her particular job function and responsibilities, employees are no longer simply laborers. They become members of a partnership and acquire managerial awareness. In meeting their responsibilities, they can learn to enjoy their work and attain a sense of accomplishment. They share a common objective in working to contribute to the company, and they develop a real sense of worth in what they are doing.

From his or her particular standpoint, each employee endeavors to contribute to the amoeba and thus to the entire company. Additionally, amoeba leaders and members set their own objectives and are drawn to the challenge of achievement. All employees discover enjoyment and a sense of worth in their work, and are thus ready to apply their very best efforts. Ultimately, employees are able to raise their abilities to the maximum and to develop as human beings.

As the name would suggest, "management by all" allows all employees to take part in management and consolidates their strengths for the development of the company, also enabling them to work with a sense of purpose and achievement. This is the third objective of Amoeba Management.

2

Business Philosophy Is Crucial for Management

DIVIDE INTO SMALLER MANAGERIAL UNITS

Applying Amoeba Management to the company requires a number of essential elements. Of these, I will explain three points that are especially important for gaining a correct understanding of Amoeba Management.

Three Requirements for Dividing an Organization into Amoebas

The first of the crucial elements I will describe is probably the single most important factor determining the success or failure of Amoeba Management. It is the question of how to divide a complex organization. Dividing the organization requires that you have a thorough understanding of the realities of your operations and apply Amoeba Management based on those realities. I believe that three conditions are needed to achieve this.

The first requirement for creating amoebas is the existence of clear revenue earned by each potential amoeba, and the ability to detail the expenses needed to earn that revenue. This is necessary to establish the amoeba as a self-supporting accounting unit, which requires an accounting of income and expenses. Creating these accounts requires the amoeba to have a clear grasp of its revenue and expenses as a fiscally independent business unit. This is the first requirement for dividing into amoebas.

The second requirement is that amoebas, as small, discrete organizations, are "self-contained business units." In other words, amoebas must be units with minimal functions, operating as single, independent businesses. It is an amoeba's status as a single, independent operation that gives amoeba leaders room for inventiveness and stimulates a sense of purpose. Therefore, an amoeba must, by necessity, be a self-contained business unit.

This concept can be explained using our Ceramics Production Department as an example. The first section that we set up as an independent amoeba was the Material Preparation Unit. Preparing raw materials is the first step of the production process. When we set out to define *material preparation* as a single, independent amoeba, we began by considering how we might make it into a self-contained business unit.

At the time, I became concerned that perhaps I was subdividing the operations too much. It then occurred to me that outside suppliers were already selling fully prepared raw materials to fine ceramics makers like Kyocera. If these external companies could specialize in mixing raw materials, then an internal Kyocera group should likewise be able to obtain the unprocessed raw materials at a low cost, prepare them in-house, and sell them to the next stage of our production process—a stage called *forming*. We should therefore be able to turn "raw material preparation" into a credible independent business. On that basis, I decided to establish the Raw Material Preparation Unit as an amoeba.

In the forming process also, many small outside factories already offered that kind of service. When provided with raw materials and machines, they could be contracted solely to conduct forming work. Again, within Kyocera, the forming unit could buy the prepared raw materials, recording a purchase, and then earn revenue by selling the formed work-in-process to the sintering unit. The result would be self-supporting accounting departments. Thus, the organization could be subdivided into units with the smallest possible number of functions, allowing them to operate as independent businesses.

However, we cannot assume a smaller amoeba is always necessarily better. Subdividing to excess can inundate a company with small organizations and generate waste. In Amoeba Management, as just discussed, the first condition is to have clear revenue and expenses passing between amoebas. Therefore, setting transaction prices between amoebas, and determining how to resolve potential issues such as quality problems, can make the task of management very complex.

In addition, even in a small organization, amoeba leaders need to feel that they are doing something worthwhile as managers. An organization should not be subdivided to the point that leaders become unwilling or unable to innovate and improve their operations. Thus, the second requirement of creating amoebas is subdividing the organization into units that can stand as individual enterprises.

The third requirement is that subdividing an operation into amoebas must not hinder the objectives or policies of the company as a whole. Occasionally, subdividing into amoebas may create internal disharmony to a degree that could actually impede the company's overall mission. Therefore, even if the revenue and expenses of a certain organization can be clearly captured as a self-contained business unit, we should make that organization an amoeba only if we are convinced that doing so would support the company's overall mission in a meaningful way.

Let me illustrate this with an example from the Sales Department. When a custom-order manufacturer grows, like Kyocera, the Sales Department is divided: the sales section visits clients to obtain orders, the delivery control section handles production scheduling and other matters relating to the delivery of products, and the accounts receivable section deals with the issues of invoicing and collections. It may be possible to apply self-supporting accounting to some sections. Suppose the Sales Department as a whole is entitled to a commission of 10 percent of sales, which represents its revenue. The sales section could then take 5 percent, the delivery control section could receive 3 percent, and the remaining 2 percent could go to the accounts receivable section. Self-supporting accounting could be applied to each section.

In such a case, though, it may not be possible to provide clients with consistent service as a business. For instance, when dealing with major customers such as Company A, B, or C, it is not enough for the Sales Department to simply accept orders. There are delivery schedules to be managed, products to be delivered, immediate responses needed in case of a customer complaint, and accounts to be collected. If each of these functions were undertaken by separate sales amoebas, then Kyocera would be unable to provide clients with consistent service. Because we would be unable to do business in accordance with the "customer first" principle of our company policy, the Sales Department is not divided into the smallest possible individual units.

As this example shows, just because it is possible to divide an organization into ever-smaller amoebas, doing so is not necessarily a good thing. The organization should not be divided beyond the level at which all units can still uphold the policy of the company as a whole. Again, this is the third requirement of creating amoebas.

Only when these three requirements have been met does it become possible to create truly independent and functional amoebas. At Kyocera, it is said that "The manner of dividing amoeba organizations is the *be all and end all* of Amoeba Management." This is no exaggeration. How the amoeba organizations are formed is a crucial element of a successful Amoeba Management system.

Continual Review of the Organization

Having divided an organization into amoebas, is the work finished? Absolutely not. A feature of Amoeba Management is the ability to flexibly rearrange amoebas in prompt response to sudden changes in economic conditions, the market, technological trends, competition, and other factors. Our operating environment changes moment by moment. To succeed, we must stay in the best possible position to respond to market changes and new competitive forces as soon as they emerge. Our leaders must constantly monitor whether the organization remains best suited for the current business environment and the objectives and policy of the enterprise.

Some time back Mr. Kensuke Itoh, then the president, initiated a new operating division, the Logistics Service Division. The Logistics Service Division was set up as an independent, profit-earning business some 30 years after Kyocera's founding. Before then, Kyocera's shipments were consigned to outside carriers by the business systems administration departments of individual factories; but since delivery is an industry by itself, we decided to organize all in-house deliveries and distribution activities as an independent division. Divisional profit began to climb almost immediately as freight costs dropped sharply. Each factory was expected to maintain close watch over freight expenses, but still the newly formed Logistics Service Division revealed waste.

There may well be other in-house operations that could benefit from becoming independent amoebas. Top management must constantly review the entire organization from the perspective of operating efficiency. Existing amoebas must be constantly reviewed to ascertain whether conditions are right to subdivide further or, conversely, whether reintegrating amoebas might perhaps produce efficiency gains. Maintaining an organization of amoebas at optimal efficiency is a task of extreme importance. Failure to do so could cancel out any benefits of introducing Amoeba Management. This is why we say, "The manner of dividing amoeba organizations is the *be all and end all* of Amoeba Management." Based on the three conditions defined in this section, it is crucial to continually examine whether the organization is optimal for the nature of our operations at any given time.

DETERMINING INTER-AMOEBA PRICES

In manufacturing, creating amoeba organizations at each step of the process facilitates the purchase and sale of work-in-process between amoebas. Selling prices must then be set for inter-amoeba transactions. Because each amoeba is inclined to maximize its own profits, the question of "how to set the selling price" becomes an important concern for the amoeba leader.

To determine the selling price between each processing step, we start with the price paid by the company's end customer and move backward. Consider, for example, a ceramic product that passes through Raw Material Preparation, Forming, Sintering, and Processing before completion. Starting with the booking value, the selling price between amoebas is then allotted to each processing unit in turn, starting with the final stage (Processing) and moving backward to each of the prior stages: Sintering, Forming, and, finally, Raw Material Preparation. However, caution is needed, because the order price is the only objective basis for determining purchase and sales prices between amoebas.

A Fair Judgment

How, then, is the price between amoebas to be determined? The price between the processes is determined by working backward from the selling price to the customer. When the price is set, inter-amoeba selling prices are determined according to the principle of using roughly the same rate of hourly efficiency for all processes needed to make the product. The product is sold to the customer for a certain price. Let me repeat: the inter-amoeba purchase and sales prices are determined for the final stage (Processing) and then backward from there to establish prices for Sintering, Forming, and Raw Material Preparation.

If one department attains a high selling price and secures sufficient profit, disputes can arise. Another department may claim it is unable to make a profit, no matter how hard it works, because its selling price is too low. Such disputes and notions of unfairness between amoebas must be avoided. To prevent this from happening, top management, the final decision maker on prices, must be recognized by all concerned as having set fair prices. The individuals who decide inter-amoeba selling prices must set prices fairly. To do so, they must have a thorough understanding of the expenses generated by each department, how much labor is needed, the level of technological difficulty of the product, and comparative market prices for similar products. In other words, the individuals who set inter-amoeba selling prices must be fair and impartial, and have the discernment to persuade all parties of the rationality of their decisions.

Furthermore, to make fair pricing decisions, top management must also have adequate societal common sense regarding the value of labor. By *societal common sense*, I mean generally accepted awareness, in this case of the value of labor. For example, what percentage gross margin is required when selling specific electronic devices? What is the hourly rate of payment for undertaking a particular job part-time? What wages are paid for subcontracting a certain process? Top management must remain well informed on such matters at all times.

The following example highlights the need for such knowledge. Let's say a company decides to produce a high-tech device with high added value. The production process has many steps requiring highly advanced technology. Among these steps we have Amoeba A, which has responsibility mainly for one routine task. Using our price-setting principles, inter-amoeba prices are determined on the assumption that all amoebas have the same rate of hourly efficiency. Because the product has high added value, all processes are allotted selling prices at the same high hourly efficiency rate.

Accordingly, Amoeba A, which undertakes much simpler, routine work, is also allotted a high hourly rate. In comparing that rate with the cost of subcontracting the work outside the company, it turns out that Amoeba A is getting a price much higher than the market rate. Paying Amoeba A for work at a rate many times its value on the open market would amount to revenue gained without effort. In contrast, Amoeba B, at another processing step, requires high technological capability and is engaging in ongoing capital investment. Due to the various work-related expenses, Amoeba B should rightly receive a higher proportion of added value. In this case, top management needs to exercise societal common sense to ensure Amoeba A does not receive excessive profit; we must set Amoeba A's selling price at a level in line with the open market.

Armed with a good understanding of the work undertaken by each amoeba, top management must evaluate the costs and labor incurred by amoebas in terms of societal common sense. Top management must set—and be recognized as having set—correspondingly fair selling prices.

LEADERS MUST: MANAGEMENT PHILOSOPHY

Conflicts of Interest Destroy Morale and Profits throughout the Company

Although top management may set inter-amoeba selling prices that are fair in the light of societal common sense, conflicts of interest and disputes between amoebas can still occur.

Let us assume that there is a product for which fair inter-amoeba prices have been initially determined. Two months later, competition with other companies in the same industry has caused the product's market price to fall 10 percent. If all inter-amoeba selling prices can be uniformly lowered 10 percent, all will be well. However, because amoebas operate independently, conditions are different in each. One amoeba might declare, "We accepted the initial selling price knowing it would be tough. Now we are being asked to accept a further 10 percent drop in price, which will undermine profit and remove any value in producing the item. We don't need this order." With that, a uniform drop of 10 percent becomes problematic.

Assuming responsibility for managing their respective amoebas, leaders coordinate on inter-amoeba prices. They cannot simply accept discounts that will result in serious deterioration of profits. Trying to minimize the burden of price reductions, they argue from their own standpoints, resulting sometimes in vehement disputes.

Under Amoeba Management, each amoeba leader strives to improve profits ultimately for the benefit of all. Therefore, in the process of eking out a little more profit, there is a tendency toward egoism. However, disregarding the standpoints of others in a drive to maximize one's own profit can have serious consequences. The effect on in-house personal relations is akin to throwing gravel into the gears of a machine.

Similar confrontations can emerge between sales and production departments. Many manufacturers apply a sales method known as the *sell out and buy up method* for transactions between Production and Sales.

Sales departments set out to make a profit by obtaining goods from production departments as inexpensively as possible, and selling to customers at the highest possible prices. There is the appeal of applying one's talents to doing business, like an independent trading company.

Yet, in the case of direct selling undertaken by such manufacturers as Kyocera, the sell out and buy up method would have a ruinous effect on profit for the company as a whole. Sales departments would try to buy as cheaply as possible, and production departments would try to sell as high as possible, resulting in conflict between Sales and Production. If one or the other gained because of egoism, the conflict would be intensified. This could conceivably impoverish the company.

To prevent such an outcome and block conflict between Sales and Production, I introduced a commission system. Sales departments automatically receive a commission of, say, 10 percent of sales. With such a business model, sales departments cannot make a profit as a result of price-haggling talent alone. Instead, as long as they produce sales, they automatically receive a fixed percentage as commission.

However, with this model, because the Sales Department receives a fixed percentage of sales even if the selling prices decline, Sales has readily agreed to customer demands for lower prices. For Production, reducing costs by any percentage is extremely difficult and a serious problem that may make the difference between profit and loss. Notwithstanding, the Sales Department sometimes easily agrees to customer demands for lower prices. This has resulted repeatedly in conflict between Production and Sales. In spite of a commission system for determining Sales Department profits, the Sales and Production Departments engaged endlessly in confrontation.

Similar confrontations also occurred between overseas sales companies and headquarters in Japan. In 1968, Kyocera opened a representative office on the west coast of the United States. The following year, we established a U.S. subsidiary company, Kyocera International Inc. (KII), which began selling fine ceramic components mainly in Silicon Valley of California. However, whenever complaints or problems with deliveries emerged, disputes quickly arose between the Kyocera production departments and the local salespeople of KII. Salespeople in the United States were irate over the notion that their sales performance was not improving because of production in Japan. In those days, communication was conducted chiefly by telex, and protests arrived by telex in Japan one after another.

In the case of customer complaints, Production and Sales should normally cooperate in an endeavor to restore customer confidence. In fact, once an internal quarrel occurred at a critical time, and came to the attention of a customer. Sales personnel had been berated numerous times by customers over delivery problems. Their undaunted response was "This is the fault of Kyocera Production. I have telexed them again and again, but Production fails to keep its promises." To save face, sales personnel criticized their company's production department to the customer. Such statements were made despite the very real possibility of destroying trust in the entire Kyocera Group and losing all future orders.

This kind of confrontation is the outcome of "egoism"—the desire to save one's own skin. However, by dividing the company into small organizations under Amoeba Management and running each operation with self-supporting accounting, the desire to maximize the profit of one's own division becomes the top priority. Departmental egoism emerges quickly, and relationships may become awkward. To rephrase: under Amoeba Management, the desire to protect one's organization becomes even stronger; this tends to intensify interdepartmental strife, which could potentially throw harmony into disarray throughout the company.

Leaders Must Be Impartial Umpires

Since each amoeba earns its own sustenance, it cannot survive without the instinct of self-protection. At the same time, from the perspective of the company as a whole, the primary mission is to maximize total profits. If there is conflict between an amoeba's profitability and that of the company as a whole, discord will ensue. To overcome this conflict, even as amoebas safeguard their own interests, they need to rise above differences in perspective. They need to consider circumstances beyond their own and make decisions from the vantage point of a higher dimension. This is where a philosophy—the philosophy of management—comes in.

I use the word *philosophy* in reference to a decision-making criterion that I have advocated continually: asking ourselves to "Do what is right as a human being." Making this universal management philosophy the backbone of management eliminates inter-amoeba clashes of egoism, and encourages harmony between amoeba profit and total-company profit. Based on this philosophy, Amoeba Management stimulates the simultaneous pursuit of amoeba and total-company profit by giving a real solution to inter-amoeba conflicts of interest. In other words, only when it is based on this philosophy can Amoeba Management overcome conflicts of interest and function as intended.

Leadership candidates may include people who are very assertive and forceful. I have always said that assertiveness and, to a certain extent, even the willingness to fight for what you want are necessary qualities. However, consider an emerging conflict that leads to a dispute within the company. An obstinate leader with an abrasive attitude, one who is ready to trample over the standpoints of others to maximize the profits of a single amoeba, cannot safeguard the profits and morale of the company as a whole. For this reason, leaders must acquire an elevated philosophy to help set higher minded standards and avoid egocentric actions.

When a conflict becomes intractable, the manager above the amoeba leaders in dispute must be prepared to mediate. In such a case, this manager needs to listen carefully to the assertions of both parties. This individual must make truly impartial judgments, and ensure that both parties are ready to abide by his or her decision. In the case of internal transactions, the mediator must be able to make a fair judgment, such as "On this particular point, you are mistaken. You need to put in a little more effort and lower the price."

"Do Not Lie," "Do Not Cheat," and "Be Honest"

Lately, large corporations have been caught up in one scandal after another, including fraudulent practices such as filing false reports and embellishing accounts to present the company's performance as better than it actually is. Fearing that truthful reporting would put them at a disadvantage, some companies falsify reporting data or engage in other dishonest practices. In many companies, management makes no effort to maintain ethics. To varying degrees, such companies exist in Europe, the United States, and elsewhere, not just in Japan. Wrongdoing emerges as an organization grows. If we could look inside the world's largest organizations, we might discover that many of them are already in an advanced state of decay.

Such problems stem from the fact that leaders and senior management lack the self-discipline and ethical values needed to prevent selfish activities. These ethical values are not part of any advanced philosophy. They are the same basic ethics taught to grade-school children: "Do not tell lies," "Do not cheat," and "Be honest."

If you told a manager who engaged in dishonest practices, "You must not tell lies or deceive people," that manager would probably agree with you 100 percent. Yet, knowing something and doing (or not doing) it are two different things. Because this intellectual knowledge has not been made part of his or her body and soul, the manager will choose deception when the opportunity arises.

In general, it is believed that expanding a business requires clever executives and top-class personnel with advanced business abilities; therefore, corporations recruit graduates from leading universities, entrusting important responsibilities to such people. However, it is often said, "A genius drowns in his talent." In other words, when top-class personnel go astray in the application of their talents, it can cause unexpected problems. Too much of a good thing can work against them if they do not know how to control their talents. Deceptive business practices are less likely to be committed by people who are unsophisticated, but they find fertile ground in people who have business talent without self-discipline.

Talent is essential for business; however, the greater the talent, the greater the potential for disaster if an individual does not have the character necessary to keep such talent within ethical bounds. A seemingly endless series of unbelievable wrongdoings has been committed by top-level managers who have succumbed to their selfish desires.

Among sophisticated businesspeople, the more talent they have, the easier it is for egoism to take hold. Human talent is basically activated by the character of the individual. To suppress egoism, we must therefore focus on raising the level of individual character. However, before delving into the matter of individual character, it is necessary to first establish primary ethics. It seems as if managers all over the world have forgotten even these fundamental ethics.

Even in our company, with its Amoeba Management, there are amoeba leaders who manipulate production accounts and otherwise try to present the operations of their units as better than they actually are. Leaders have a duty to honestly declare, "We did not do well," if, in fact, performance was poor; yet many try to smooth over the results, fearing reproach from superiors and colleagues. Such people do not have the true courage needed in a leader.

Kyocera prizes the very basic values of fairness, impartiality, justice, courage, sincerity, perseverance, effort, and philanthropy. Probably no other company in the world places as much importance on these basic values. The emphasis placed on these values makes it possible for the Kyocera Group to sustain and foster high ethical values and a superb corporate culture.

I always say that a leader must be a person of all-around excellent character. Human character is constantly changing. If people succeed and win praise, they may become proud and lose their humility. Without constant self-discipline and ongoing learning, excellent character cannot be sustained. To be able to lead groups in the right direction, leaders need more than ability and competence in their work. They must continually strive for personal development, improve and polish the mind, and endeavor to become people of wonderful character. At Kyocera, we want all managers, from our top executives to those leading the smallest amoebas, to develop a truly elevated human nature as a part of their leadership responsibility.

Embodying the Philosophy in Management

In Amoeba Management, concepts of the Kyocera philosophy are strongly reflected in its system of remuneration. Kyocera does not use a system that tries to manipulate people's minds through money. Wage increases and bonuses are not linked directly to improvements in the hourly efficiency of an amoeba. Of course, the results of work are duly evaluated and reflected over a longer period in the development of individual careers and conditions. However, there is no system for raising wages and bonuses according to increases in hourly efficiency. Instead, there is the intangible but highly rewarding honor of praise and appreciation by companions who believe in each other, for the outstanding performance of the amoeba and the contribution made to the company overall.

Claims that this system works very well may be difficult for someone outside of the company to understand. The fact is, based on the management rationale that "Our ability to contribute to the happiness of co-workers who believe in us justifies the existence of our amoeba" is a deeply rooted concept at Kyocera. Therefore, the highest honor that can be given is praise and recognition from one's peers for contributions to the company. Amoeba Management is a management system embodying the management rationale: *to provide opportunities for the material and intellectual growth of all our employees and, through our joint effort, contribute to the advancement of society and humankind.*

As I stated earlier, Amoeba Management is "management by all" based on mutual trust between managers and employees, and among the employees themselves. Because all employees are taking part in management, all employees can progress toward their respective goals, whether they work in a factory or visit customers to promote sales.

At Kyocera, the awareness in each person that "I, too, am a manager" provides a sense of worthwhile purpose in work, and the ability to share the enjoyment and appreciation of the results with others. Thus, Amoeba Management is management that allows employees to feel the enjoyment of management. In its esteem for the work of each person, Amoeba Management is "management with respect for human life and dignity."

Place Highly Capable Leaders

A vital point in running an organization is the placing of truly capable people at the top of the organization. Suppose a company used a system of paternalism, by which a person is made leader on the grounds of seniority in years. Company management would soon come to a standstill, with all employees bearing the resulting misfortune. In contrast, imagine an individual who, though lacking in experience, possesses superb character, is highly capable, has enthusiasm for work, and is respected as a person and trusted by others. Placed in the right position, such a person would be capable of taking the company through intense competition and on to growth.

As a rule, Kyocera has been run as a merit-based organization. In a merit-based organization, people with true ability, irrespective of age or tenure, are selected for positions of responsibility in which they can guide the company toward prosperity. As these selected people apply their abilities and produce results, we believe that, in the long run, they will develop and grow in the positions for which they were selected.

However, the application of merit-based organization may create a problem. If, for instance, a relatively young man with real ability and popularity is selected as a director, he may attract the anger of senior people in his circle: "That fellow is three years my junior. It's outrageous that he should be appointed as director before me!"

To that, I reply, "Instead of simply becoming angry, senior employees should stop and consider rationally whether their appointment as a director instead of the young man would truly be for the greater benefit to the company. If you stop to think about it earnestly, you too may decide that the younger fellow is perhaps indeed capable of making a greater contribution to the company as a director. Entrusting capable young people with leadership of the company will add to the happiness of all employees. Rather than begrudging the selection of younger people and holding on to resentment, this should be a cause for genuine approval. I do not want Kyocera executive personnel to think of themselves as 'next in line' because of seniority, but to be broad-minded to the extent that they simply want the company to be led by people with genuine ability."

The following episode occurred a long time ago. More than ten years had passed since the company was established, and the company shares were about to be listed. If we were to expand the business, we would need to develop new fields. That required people with various experience, technologies, and knowledge. I was considering bringing in people from outside of the company to take on this daunting task, and discussed the matter among the executives with whom I had been running the company since the start.

"The fact is, I am thinking about bringing this person into the company. I am considering placing this person in a position of authority higher than yourselves, with whom I have worked since the company was founded. What do you think? If you tell me we have worked together building this company and that you do not want someone you know nothing about coming in and standing above you, then I will abandon the idea. However, think about the expression 'A crab digs a hole that suits its shell.' Apply this to us, and note that the company cannot grow any larger than the capabilities of the management team. If you tell me that Kyocera was not established with such a narrow-minded outlook, that we pledged to build this company into the world's number-one enterprise, and that you have no objections to hiring midcareer executives to these higher positions of authority, then I will go ahead with my plan."

Having heard this, everyone very agreeably consented to my proposal, and I brought the person to our company. Before long, no one could deny that the top-class personnel who were hired midcareer contributed substantially to the growth of Kyocera. The people who had taken part in founding Kyocera understood that the merit-based organization is the basis of company development. They clearly saw that the merit-based organization is the means of bringing wonderful benefits to all employees. This episode clearly illustrates the origins of Kyocera's merit-based organization.

Within this concept, Kyocera has also endeavored to acquire people from outside of the company who have both real ability and a pleasing character. Whether in regular hiring sessions or in midcareer hiring, with no reference to academic cliques or "old boy" networks whatsoever, we have sought to hire people, young or not so young, who have ability and agreeable personalities. The merit system is an important principle of running an organization under Amoeba Management, and it is a management principle that has sustained the growth of Kyocera.

Performance-Based Pay System and Human Psychology

Whereas Kyocera management is based on the merit-based organization, many European and American companies employ a performance-based pay system. In the Euro-American method of performance-based pay systems, incentives increase or decrease substantially depending on work yield. This method appeals directly to the worldly desires of employees. It is an impersonal system that allows for bonuses in the case of high work yields, no bonuses if yields fall or fail to rise, and in some cases even dismissal.

I have long believed that a manager needs good insight into human psychology. Performance-based pay systems raise employee motivation with the promise of large rewards if yields rise. In the short term, this may be an effective management method. However, yields cannot rise forever. Inevitably, the time will come when yields fall. The human mind is a mysterious entity, and unintentionally it becomes accustomed to receiving large rewards due to rising yields. Therefore, when yields deteriorate and incentives fall, few people are rational enough to see that times had been good and accept the decline with equanimity. Morale falls, and dissatisfaction with the company begins to build. Under such circumstances, company operations cannot hope to proceed smoothly.

Furthermore, some companies base incentives on what they call *payment by results*. Incentives for each section rise or fall according to the results they produce. In consequence, morale will rise in sections with good performance and fall in sections with poor performance. A mentality of begrudging resentment will inevitably emerge.

With performance-based pay systems, incentives fall if performance deteriorates. Many employees subsequently develop feelings of dissatisfaction and resentment. In the long term, this can wreak havoc with the company morale.

In particular, Japan is culturally homogeneous and possesses a strong consciousness of equality, resulting in considerable psychological resistance to large differences in remuneration and job conditions. If you apply the Euro-American style of straight performance-based pay systems in Japanese enterprises, at first the organization will appear to come alive with an air of "We'll get good bonuses if we make the effort." After a number of years, however, this system will surely invite a disruption of company morale and foster resentment.

That is not to say that all employees should receive the same job conditions. If conditions for people who work hard for the good of everyone else are the same as for those who do not, then that in itself amounts to inequality. Under Amoeba Management, there is little difference in individual incentives based on short-term yields. However, for individuals who work hard for the benefit of everyone else and show good results in the long term, ability is duly assessed and reflected over the longer term in wage increases, employee bonuses, promotions, and so on.

Turn All Operations into Businesses That Cannot Be Imitated

When people talk about running a manufacturing company and other enterprise, they say that those possessing advanced technology or high-tech capabilities are outstanding. Companies holding strong patents, advanced technology, and so on are acclaimed as outstanding enterprises.

Of course, patents and advanced technology are important. Yet, even if these patents and advanced technologies currently lead the market, competitors will devise and execute new methods within a few years. If a company relies solely on technology, what happens when another company catches up with it? In today's world of fast technological change, making good use of high-tech capabilities and developing new technologies one after another are extremely difficult. Therefore, when the competitor catches up, the initial advantage is eliminated at a stroke.

A technical advantage is not permanent. What, then, can we do to stabilize our corporate business, even though we do not lead the way technologically? The key is to have outstanding operations that can be undertaken by any number of companies. In other words, even when performing work that anyone can do, the true ability of a company is realized when people can say, "That company is different."

Recently, some of Japan's top general electronics manufacturers have taken note of the brave fight put up in the face of poor business conditions by Rohm, Murata Mfg., Kyocera, and other leading electronic component makers based in Kyoto. They declared, for example, "We cannot make profit through assembly. Perhaps we should turn our attention to manufacturing electronic components and devices."

They say there is no profit in assembly. I disagree. The work of assembly is not simply a matter of installing parts onto printed circuit boards. This superb business entails designing circuits, combining components, and adding outstanding functions before delivering the finished product to the customer.

Combining components in such a way that the functions exponentially raise the value of the product is the primary job of the assembly manufacturer. If such techniques are properly applied, then assembly should be able to achieve good profits. If the company forgets its original purpose and decides to go into manufacturing electronic components because it is currently not making a profit in assembly, then it cannot possibly be successful in a new venture. Whatever the field, wisdom and effort applied unsparingly to the creation of new products that catch the minds of customers will open the way to unlimited added value.

This is not limited to high-tech manufacturing. Many of the small and medium-sized enterprises (SMEs) that support Japan's economy make garments, towels, shoes, and other ordinary items. More than a few such SMEs have been driven into bankruptcy by cheap Chinese imports or for other reasons. In contrast, there are also companies that enjoy thriving operations due to single-minded creative endeavors and determined efforts that cannot be beaten.

These companies produce fine results in industries that have been around for a long time. They do not stand out in the business world. Yet, they are truly uncommon companies in that they are doing a wonderful job with ordinary work.

KCCS Management Consulting (KCMC), a Kyocera affiliate, undertakes Amoeba Management consulting. Many of the enterprises that have introduced Amoeba Management are operating in unglamorous, long-standing industry sectors. However, with the introduction of Amoeba Management, these enterprises are raising employee awareness of participation in management, and implementing profit management right down to the individual amoeba units. As a result, they are continually raising the added value of their work and increasing profits without bounds.

Therefore, it is possible to achieve high-profit business doing ordinary work in the absence of cutting-edge technology. Even a plain, ordinary business can shine. This is the true worth of Amoeba Management.

3

How to Organize Amoebas

SUBDIVIDE THE ORGANIZATION AND CLARIFY EACH FUNCTION

Functional Needs Are a Prerequisite for Forming Departments

The organization is an extremely important part of the foundation of company management; organization formation becomes a central pillar of management. Many companies in general follow the so-called common sense of management when they form their organizations; however, simply building organizations according to common sense makes it easy for the number of employees to grow and for the organization to expand beyond what is necessary.

For example, in the case of a recently established manufacturer, following the common sense theory of general organization means that the company needs administrative departments for handling accounting, personnel, general affairs, and purchasing, in addition to production, R&D, and sales. If the departments are then divided into sections with people responsible for each section, then the number of organizations grows, as does the number of necessary personnel.

To avoid this kind of organizational expansion, enterprises must be structured according to the functions essential for running the company. Avoid the concept of "keeping pace with the neighbors" and creating various organizations just because other companies are doing so. Efficient administration of a company begins with clarification of the functions that are necessary, followed by consideration of the absolute minimum number of departments needed to perform those functions and then of the minimum number of necessary personnel.

As an example, when Kyocera was established, we did not set up individual departments for handling accounting, personnel, general affairs, purchasing, and other matters. The reason was that, as a manufacturer, we were unable to spare many people for work outside the minimum necessary functions of production, R&D, and marketing. We set up just one administrative division for dealing with all other kinds of work. As a result, we were able to form a lean organization with absolutely no waste, in which just a handful of people dealt with all the work outside of production and R&D.

Based on the idea that "Functional needs are a prerequisite for forming departments," Amoeba Management proposes the construction of lean organizations that fulfill the minimal necessary functions with no waste.

Each Member of the Organization Should Have a Sense of Mission

For some time after Kyocera was established, I was personally involved in all aspects of business operations. I accepted orders from customers, developed products, and undertook to manufacture them. From that experience, I understood that running a manufacturing company has four basic and minimal requirements: sales, production, R&D, and administration. Therefore, I set up the departments shown in Figure 3.1.

Many manufacturers today have very function-specific departments. However, simply dividing organizations according to function is not enough. For a company to conduct operations as a unified body, it is vital that employees belonging to each department possess a thorough working knowledge of their department's functions or roles and are imbued with a sense of mission to fulfill those functions.

For example, the role of Sales is to seek and accept orders from customers, to secure work for Production, to provide satisfactory products and services to customers, and to collect payment. Meanwhile, Production aims to consistently produce goods that satisfy customers in terms of price, quality, service, and delivery schedule, thereby creating profit. The role of Production is thus to make outstanding products while cutting costs to an

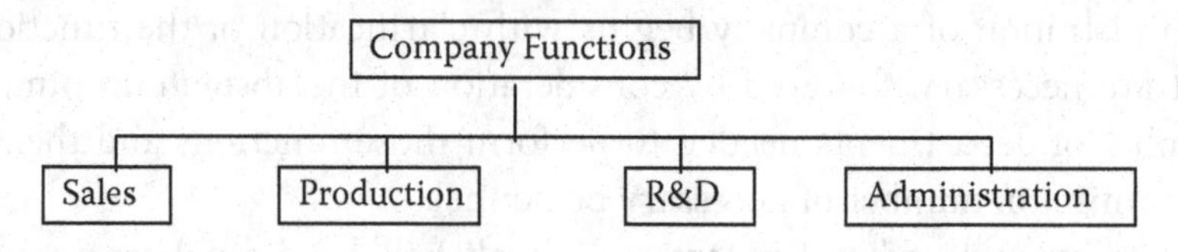

FIGURE 3.1
Kyocera's organization by function in the early days.

Function	Basic Roles
Production	Create added value through the manufacture of products that satisfy customers.
Sales	Raise the level of customer satisfaction and added value through sales activities (from the acceptance of orders through to payment collection).
R&D	Develop new products and technologies based on market needs.
Administration	Support the business activities of each amoeba and ensure the smooth running of the company overall.

FIGURE 3.2
Basic roles of Production, Sales, R&D, and Administration at Kyocera.

absolute minimum, and to work constantly at raising the level of added value. The basic roles of Production, Sales, R&D, and Administration at Kyocera are given in Figure 3.2.

Furthermore, businesses inevitably have workflow patterns that consist of many processes. Operations cannot advance smoothly unless each of these processes fulfills its necessary functions in cooperation with other processes. When creating new organizations, this kind of business flow must be clarified. The necessary functions of each process need to be clearly defined, and each function must be fulfilled in the course of business flow.

For a company to use its organizational strength, each person in each department must have a deep awareness of his or her roles and responsibilities. For each person, a strong sense of mission to fulfill those roles and responsibilities is essential. Although this may appear obvious, it is nonetheless the most important element of the ideal state for Amoeba Management.

The Subdivision of the Organization Must Satisfy These Three Conditions

How should we subdivide departments that have been categorized according to functions, and build amoeba organizations?

While bearing responsibility for one of the functions performed by the company, each amoeba is a single unit operating independently under self-supporting accounting. Therefore, it is not simply a matter of subdividing the organization. How the amoebas are formed is a vital issue to the extent that it determines the success or failure of Amoeba Management.

As mentioned in Chapter 2, there are three necessary conditions for the formation of amoeba units:

1. An amoeba must be a fiscally independent unit. That is, an amoeba must have clearly definable revenue and expenses.
2. Amoebas must be self-contained business units. In other words, for a leader to manage an amoeba, there must be room for creative inventiveness and the ability to maintain a sense of worthwhile purpose in conducting operations.
3. The subdivision of the organization must support the company's goals and objectives. This means that dividing the organization into amoebas should support, not hinder, the accomplishment of the company's goals and objectives.

When I began dividing the organization, I began by focusing on the Production Department, which had a major impact on the company's profits. At the time, we were manufacturing fine ceramic components in their entirety for the electronics industry. To clearly define the profits of the separate processing steps, I formed amoebas by dividing Production according to processing steps, using small numbers of people. I appointed a leader for each amoeba and entrusted that person with the overall management. In the end, I had divided the Production Department according to separate processes, as shown in Figure 3.3, and set up process-specific unit operations.

As the company grew, the number of product types rapidly increased. It then became necessary to further subdivide these amoebas according to specific products. Moreover, as the factory became cramped, we built new factories in Shiga and elsewhere. The creation of factory-specific departments became necessary. In this manner, and in line with the company's growth, the number of amoebas rapidly increased with the formation of

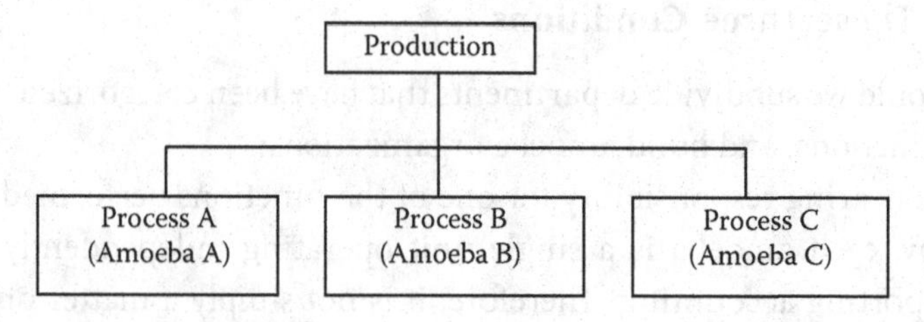

FIGURE 3.3
A subdivided organization by process in the manufacturing division.

different process-specific, product-specific, factory-specific, and other types of departments.

At the same time, organizations were further subdivided as the Sales Department was broken down into amoebas for particular regions, products, clients, and so on. Similarly, the trend was extended to the R&D Department and the Administration Department.

Before long, I was planning the startup of various new businesses, aiming to stabilize operations and expand the company. I adopted a divisional system for appropriate management of the various businesses and actively promoted diversification. The result of all of this subdividing is that, today, Kyocera has approximately three thousand amoebas.

Form Amoebas Where Overall Operations Are Apparent to Leaders

What benefits does organizational subdivision bring, from the perspective of company management? Let me use the example of grocery stores.

As an example, consider a family-run neighborhood grocery store in a small village somewhere. The store, which is not very large, sells vegetables, fish, and meat. It also handles dried foods, canned goods, instant foods, and other processed foods. When a customer makes a purchase, the storekeeper puts the money in a cash basket hanging at the front of the store and hands the change to the customer. When business is over for the day, the storekeeper tips out the contents of the basket and counts the day's sales. This grocer employs the so-called *rough-estimate* style of business management.

Such a method is probably adequate if the store handles only a limited number of product types and the storekeeper relies on experience and business intuition. However, if, as in this case, the store handles many different product types, obtaining a true picture of operations is difficult. Because the nature of products differs with product type, the business methods should also differ.

First, there are considerable differences in product life for different items. Meats and frozen goods will keep for a while in the refrigerator. Fish will probably not keep for more than a day, and vegetables will soon wither. As product life differs, price setting for different product types will also differ. Goods that are disposed of if not sold must, to an extent, secure a high gross margin. Goods that have a long shelf life and sell well can be sold with a smaller profit margin. The business method changes with the nature of the product, and each product type must be managed separately. Department-specific accounting should be introduced to clarify which sections are creating profit and which are not.

Using the grocery store example, we calculate profit separately for four sections: vegetables, fish, meat, and processed foods. If totaling the sales and expenses for each section is difficult, then hang four cash baskets to hold sales revenue. Money from the sale of vegetables goes into the basket for the vegetable section, and so on, allowing for department-specific calculation. The main point is to divide the one basket into four. After trading hours, the day's sales for each section are simply counted from the separate baskets.

Under no circumstances should expenses for stock purchases or other general costs for the shop be taken from the money in the baskets. Such expenses are paid from a stockpile managed separately from the baskets, and the amount taken from the stockpile must always be specified on a voucher, in a ledger, or elsewhere. This enables the calculation of section-specific stock purchases and other expenses.

Subtracting purchases for each section from the day's sales in the appropriate basket and accounting for the value of unsold stock to be thrown out result in the clarification of revenue and expenses for each section. The storekeeper can now quickly calculate the day's net profit separately for the vegetables, fish, meat, and processed foods sections.

Examining the department-specific profits, the storekeeper might say, "I always thought we were making the most profit from vegetables, when in fact the profit from fish is highest. We need to think about how we sell our vegetables." In this manner, the storekeeper gains an understanding of the realities and problem areas of his operation and becomes able to implement prompt countermeasures.

Additionally, the storekeeper can further fine-tune operations by delegating responsibility, for instance by putting one of his children in charge of the vegetable section and another in charge of the fish. Entrusted with the management of these sections, those with less experience are in a position to devise better methods for their respective sections. In the case of the vegetables, the person in charge becomes better able to estimate how much of which vegetable is likely to sell and orders stock accordingly. He or she also begins to water the vegetables from time to time to slow down withering. Toward the evening, the unsold vegetables can be sold at a discount. This system allows the emergence of creative endeavors to make a profit even from unprofitable sections.

While this simplified example does not consider inventory, it shows that proper application of a divisional accounting system brings into view certain realities of the operation that were not visible when using the rough-estimate style of accounting. It highlights problem areas: which sections need improvement and which sections need more attention.

In other words, when subdividing an organization, a key point is determining what kinds of subdivisions will give the manager a clearer view of the true circumstances of the operation. A manager should be like a ship's captain, who has a bird's-eye view of the movement of the entire ship while steering. The important point is thus forming departments that allow a real sense of the true conditions of each department or sector, from the perspective of the manager.

Assign Inexperienced People as Leaders and Train Them

Incidentally, when breaking down an organization into smaller units, a need for leaders capable of managing the resulting amoebas emerges naturally. This brings up the issue of how to select these leaders. The problem of who to appoint as a leader becomes particularly acute when one is steadily subdividing an organization amidst a shortage of suitable human resources.

It is not necessary, of course, to push ahead with a subdivision in the absence of suitable leaders. If there is a serious shortage of people suitable for leadership positions, it may become necessary to subdivide the organization within the managerial capacity of currently available resources. It may also be possible for upper-level managers and other amoeba leaders to hold double positions temporarily after subdivision.

However, one of the objectives of Amoeba Management is the cultivation of managerial awareness in personnel. An employee with potential as a leader may be lacking in experience and ability. It's important to entrust amoeba leadership to that person. In such a case, the leader would not be left entirely to his or her own devices in managing the amoeba. The new leader would require guidance and supervision from a person capable of cultivating leadership qualities and helping him or her through areas where ability is lacking. If there is a shortage of personnel who can be fully entrusted with the leadership of an amoeba, after dividing the organization as required, you will need to select and cultivate personnel showing leadership potential.

When starting a new operation, I have always believed that "People are the source of business." Therefore, we have never started an operation simply because the market offered an opportunity. We have always begun new operations after affirming the presence of either in-house personnel suitable for taking responsibility for the new operation, or suitable outside candidates who are willing to join the company. I have a fixed rule: "Advance new projects only when there are appropriate human resources."

Under Amoeba Management, the company is divided into separate amoeba organizations. Even if things do not go well after the appointment of a new leader who has potential but little experience, there is little risk of unsettling the framework of the entire company. Therefore, it is important to allow personnel the experience and opportunity to develop managerial awareness, even if it means appointing inexperienced people who still generate uneasiness.

Organizational Subdivisions Expand Business

Kyocera's Component Operations Division was, at one stage, divided simply into Production and Sales. Before long, the Production Department was subdivided into Fine Ceramic (FC) Components, Semiconductor Components, Electronic Components, and other sections, whereas Sales was managed by just one organization, the Fine Ceramic Sales Department. In the sales offices, basically one salesperson handled the products of the one specific Production Department assigned to him or her, but when the number of salespeople was limited, as at some of the regional sales offices, one salesperson handled the products of all Production Departments.

For example, there might be separate salespeople responsible for the Corporate FC Group, the Corporate Semiconductor Components Group, and the Corporate Electronic Components Group. Yet, these three salespeople visit the same customer separately. The customer might think it would be much easier to do business with one salesperson who has responsibility for all corporate groups, instead of meeting with three salespeople. Certainly, in terms of operational efficiency, it seems more efficient to have salespeople who deal with products from all corporate groups.

However, if all sales responsibility is handed to one person who is able to accept orders for products from any corporate groups, the natural tendency will be to concentrate on the areas that are most likely to generate orders. The outcome is that insufficient time and effort are applied to cultivating new clients and new markets.

That said, the volume of orders might not be high enough to justify appointing a full-time salesperson for each corporate group. Still, each corporate group has adopted self-supporting accounting, and without sufficient orders it is not possible to continue operations.

Should we then appoint sales personnel separately for each corporate group or give priority to operational efficiency by giving salespeople responsibility for multiple production departments? In this case, the decision becomes extremely difficult. However, if we consider operational efficiency alone, then orders may stagnate at a certain level indefinitely. Even if the current order volume is small, for example, it is better to appoint full-time sales personnel and thereby raise the probability of securing larger orders in the future. Bear in mind that one of the original aims of Amoeba Management is to increase each unit's sales and independently lift profits. In that case, it is better to divide sales, even if it appears to be wasteful at present.

When organizational subdivision is possible, a degree of additional expenditure may be incurred. This should be weighed against the far greater gains that can be obtained through increased orders, sales, profits, and expansion of operations.

A FLEXIBLE ORGANIZATION ADAPTS TO MARKETS

Establish an Organization That Is Well Prepared for Action

Various points should be noted in sustaining and managing the amoebas thus formed. First, one of the objectives of Amoeba Management is the achievement of *market-oriented divisional accounting systems.* To reach this goal, in addition to subdividing the company, these amoebas must constantly adapt to changes in market conditions.

Since its foundation, Kyocera has mostly practiced "custom-order" production, which is the production of goods upon receipt of customer orders. To maintain this practice we had to think carefully about establishing flexible and efficient production systems using limited human resources and facilities to enable adaptation to market trends (i.e., adaptation to the volume and nature of orders received).

One day, the person in charge of a certain unit came to talk about his desire to "change his amoeba in a particular way, starting with the next fiscal year." My immediate response was "You, as unit manager, have observed your amoeba and noticed problems. Why, then, are you putting off reorganization until the next fiscal year? Start next month. Start now. Don't delay."

We are trying to do business in a market undergoing bewildering change. The organizational system must be flexible enough to change in tune with market movements and shifts, or the market will leave us behind. Unless we develop a system capable of doing battle today, we will be defeated in competition. This sense of crisis has pushed me to continually reform and modify the organization.

As you well know, when actually running an operation, after careful thought, you may decide to implement certain changes to the organization. However, when you think again, you decide those changes would be contradictory to the company's goals and are unlikely to go well. Some people might be concerned that this seemingly casual changing of objectives undermines their authority. However, when earnestly thinking about work, *chourei bokai*, or "Making a decision in the morning and changing it in the evening," is sometimes unavoidable.

A feature of Amoeba Management is that the workplace is highly responsive to the leader's will. If you think, "This is a good way," then follow through quickly. You may then rethink and decide, "No, this way is not so good after all. It should be done that way." Go to your subordinates and say, "I'm sorry," and amend whatever needs to be amended. Amoeba Management offers the advantage of being able to implement a wonderful idea and obtain feedback quickly. When reorganizing, work with the assumption that *chourei bokai* is necessary and orient the mind toward dynamic business development.

When practicing Amoeba Management, the organization must not become rigid. You must constantly consider whether the present organization is suitable for the current realities of the market and implement flexible reformation of the organization as needed. Even now, Kyocera organizations are continually changing and evolving according to market trends, from integration and separation of operational divisions and other large organizations on a company-wide scale down to the individual amoeba units at work floors. Therefore, the organization tables containing the names of all employees are changed each month and distributed to the management. With the organization table and amoeba-specific hourly efficiency reports, the management has a clear grasp of changes and conditions, while linking these with the faces and names of people in their departments.

The Leader Is the Manager of the Amoeba

Each amoeba unit always has a leader who is accountable for the amoeba. The leader possesses the same sense of responsibility and sense of mission that the president of a small or medium-sized enterprise (SME) would. He or she sets goals for the amoeba and undertakes management. Meanwhile, the employees regard the workplace as their own and have the awareness and desire for improvement. Each amoeba endeavors to raise its profits and improve operations. When added together, all of this amounts to major performance improvement on a company-wide scale.

Moreover, through profit management on the amoeba-unit level, top management gains an overview of movements in every corner of the company, and, thus, gives accurate direction and guidance on the management of amoeba operations. The implementation of Amoeba Management vitalizes even the farthest and smallest amoeba organizations, raising the level of the company's energy to the maximum.

Management Rational Is Crucial for Autonomous Organization

Because an amoeba engages in activities based on an independent accounting system for small groups, it may be regarded as having a high degree of freedom. Rather than simply following the commands of other divisions, amoeba members are able to do their work independently. Amoeba organizations thus raise the abilities of their members. However, as amoebas are organizations with a high degree of freedom, the level of morale of amoeba leaders and personnel, as well as their awareness of operations, are called into question.

As I wrote in Chapter 1, amoebas engage in mutual buying and selling within the company. When products flow from one process to the next, they are not exchanged at cost. They are delivered at a selling price that gives the selling amoeba a profit. However, when setting the selling price, it does not follow that the selling amoeba is free to consider only its own profit.

For instance, suppose a customer demands a price considerably lower than the final selling price for a product that has passed through numerous manufacturing processes. If we are compelled to lower the price, which units are going to absorb the discount? In dealing with this problem, amoeba leaders cannot let egoism rule. Even if profits are meager, they must agree to lower selling prices for the good of the overall operation. In other words, they must practice altruism and consider the needs and circumstances of other parties. They need to bear in mind the harmony of the entire company as they decide on their actions.

One amoeba leader might voluntarily lower his selling price: "I understand. My group will accept this price." However, in the real business world, no matter how much consideration you hold for another party, company management is not functioning effectively if lowering your selling price harms your profits. This is not true altruism. If there is genuine concern for the company, then the people involved will say, "Well, under normal circumstances we cannot make profit at this selling price, so we must devise some means of raising our profitability," and they will redouble their efforts. True altruism is the acceptance of a lower selling price based on a determination to make an all-out effort to reduce costs even further.

General managers and others might engage in price negotiations outside of the company when orders fall sharply in an economic downturn or because orders are taken by other companies in the same industry. Saying, "Market conditions are severe and it can't be helped," they sometimes give in to demands for sharply lower prices without confirming the ability to produce profit. Production responds by declaring they cannot possibly match costs at that price. The result is, in some cases, a loss.

However, for top-level negotiations, this irresponsible attitude cannot be condoned. The general manager to whom the operation has been entrusted has a responsibility to figure out how to lower costs and how profits can be secured before accepting a lower price. Additionally, he or she must be absolutely convinced that it can be done before accepting the order. The general manager will then appeal to Production, saying, "By conventional methods, we cannot make a profit. However, if we all put our energy into this new method, then there is no reason why we cannot produce profit even higher than before." It is essential that the general manager obtains the cooperation of all involved, which will ensure a united effort when overcoming difficulties.

Within the company, each amoeba is a member of a community held together by destiny, working together on the foundation of a shared management rationale. Therefore, while clarifying his or her standpoint, each amoeba leader must put aside egoism, consider the good of the entire company, and decide what is right as a human being. Moreover, with the general manager as the nucleus, all amoebas must focus on raising the profits of their respective operations, even as they retain a sense of unification with the whole.

Because a shared, universally applicable management rationale and sense of value permeate the foundations of amoeba units, the entire company can still function as one living organism even when subdivided.

BUSINESS SYSTEMS ADMINISTRATION SUPPORTS AMOEBA MANAGEMENT

As I noted at the start of this book, Amoeba Management is a business management method that I developed over the course of running a company to achieve the Kyocera management rationale. The maintenance, management, and further development of the concepts, methods, and mechanisms forming the basis of Amoeba Management are the roles and responsibilities of the Business Systems Administration Department.

The Business Systems Administration Department handles the numerical information of the entire company. Its roles and responsibilities include the accurate compilation of management information necessary to steer operations. In other words, this department is at the foundation of Amoeba Management, dealing with the correct application of management information in the same way as an aircraft navigator reads and interprets instrument panels and gauges in a cockpit.

Therefore, the Business Systems Administration Department must have a sense of mission and responsibility as the department entrusted with the application of Kyocera's management concepts: the Kyocera philosophy and Kyocera accounting principles. That is, the department must hold fast to the criterion of "Seeking the essence of matters according to truths and principles" and, thus, decide "What is right as a human being."

Outlined here are three basic roles that should be fulfilled by the Business Systems Administration Department for Amoeba Management.

Role 1: Creating the Necessary Infrastructure for the Proper Functioning of Amoeba Management

To achieve the smooth progress of business and the proper functioning of Amoeba Management, the Business Systems Administration Department constructs and ensures proper application of in-house business systems, such as the *custom-order system* and the *stock sales business system.*

In addition, the department undertakes thorough planning and revision of the in-house rules necessary for management control. When planning in-house rules and managing their implementation, clarification of the significance and objectives of the rules is essential. Please refer to the following list for a discussion on what in-house rules should be.

1. *Rules should conform to the Kyocera philosophy.* A prerequisite of in-house rules is conformity with the company's fundamental way of thinking and sense of values, as embodied in the Kyocera philosophy. Correct decision-making criteria are necessary to guide company operations to long-term success. The development of operations must be supported by in-house rules that reflect the Kyocera philosophy as a management philosophy that is shared throughout the company, as well as by policies set by top management.
2. *Rules should be drawn from the perspective of operations.* In-house rules need to be drafted from the perspective of the company's operations. Rules should be based on a good understanding of how business is conducted, and what organizational system, as well as the roles and responsibilities of those organizations, will promote operational development. Rules must be drafted from these viewpoints.

   Therefore, it is important to draft in-house rules according to business type and organizational structure, based on detailed simulations of how to measure amoeba performance (sales, net production, expenses, and time).
3. *Rules must convey the realities of operations.* In-house rules must be set to ensure that management figures state the realities and truths of the operations. This means that rules must enable recognition of the true nature of things, simplification of complex situations, and presentation of the realities of operations in a manner that can be understood by all.
4. *Rules must be consistent.* Rules must be based on consistent thinking. If you consider only a few specific cases when forming a new rule, it may end up deviating from or contradicting existing rules. Each new rule needs to be examined beforehand for its consistency with existing rules before being finalized.
5. *Rules must be fair and applicable to all group companies.* The fifth prerequisite is that it must be possible to apply in-house rules fairly and impartially to all companies within the parent group. In-house rules cannot be designed for specific business units. They must be

based on unified thinking and criteria applicable to the entire company. Individual units endeavoring to improve by learning from each other under conditions of equality comprise one of the prerequisites of Amoeba Management. To sustain this state, in-house rules must always be fair and impartial.

Role 2: Accurate, Timely Feedback of Management Information

To enable top management and each division leader to make fast and correct managerial decisions when conducting company operations, they must have an accurate and timely grasp of the current business circumstances. Like the example (given in this section) of instrument panels in an aircraft cockpit, the various management figures must reflect the current state of the business. The Business Systems Administration Department takes the lead in constructing and applying concrete methods and mechanisms designed to achieve this goal.

Role 3: Sound Management of Company Assets

In addition to a good grasp of operational results, sound management of the company's assets is vital. By company assets, I mean everything, including order backlogs, inventory, accounts receivable, and fixed assets. Accurate knowledge of company assets is an important management information tool for running the company. At Kyocera, all items and monies are managed on the principle of *one-to-one correspondence.* Results and balances are also managed to maintain conformity with the principle of one-to-one correspondence. The Business Systems Administration Department undertakes thorough management of results as well as balances. It has the role of promoting sound management and application of company assets by pushing for the proper management of assets in each department, as deemed necessary.

4

The Hourly Efficiency System: Focus on the Work Floor

ENHANCE ALL EMPLOYEES' PROFITABILITY AWARENESS: THE RATIONALE BEHIND UNIT-SPECIFIC ACCOUNTING

Use "Maximize Revenues and Minimize Expenses" to Simply Grasp Operations

This section describes the mechanisms of Kyocera's hourly efficiency system, a primary method for controlling profits in Amoeba Management.

As I wrote in Chapter 1, when the company was founded I had almost no knowledge of accounting. The first time I was shown an income statement and balance sheet, I had no clue what the columns of figures meant. Hoping to raise my level of understanding, I asked the person in charge of accounting various questions. He was surprised at the rudimentary level of my questions.

Having had no training in management or accounting, I tried to avoid thinking about business as difficult and made it as simple as possible. As a result, I came across a fundamental principle: "If we raise revenues as high as possible and reduce expenses as low as possible, then we will maximize profit, which is the difference between the two." To this day, I have run the company according to this principle.

The fundamental principle "Maximize revenues and minimize expenses" became the basis for the hourly efficiency system. To that end, reducing spending and eliminating all forms of wastefulness are fundamental elements of managing a company that provides customers with the products and services they need.

There is a tendency to accept the idea that expenses will rise as revenue increases. In fact, it need not be so. By applying intelligence and effort, it is possible to prevent expenses from rising when revenue increases, and even to reduce expenses. Raising revenue by applying inventiveness and constantly seeking to cut expenses to the bare minimum is a principle of management.

For management and all employees to apply this principle together, the issues of how revenue can be raised as well as where and how expenses are generated must be clarified in a manner that is easily understood by people at their work floors. This requires a simple, comprehensible method of profit management.

A Management Accounting Method Suitable for Work Floors

Many medium-sized, small, and very small enterprises outsource preparation of their income statements and other financial statements because they are unable to retain in-house accounting personnel. Sales invoice slips, payment slips, and other documents collected over a period of time are taken to the offices of tax accountants or certified public accountants, where they are arranged and used in preparation for the income statement. The office might send a report on the profit situation once a month if there are many documents, or once every six months if there are not many. However, this method of arriving at the numerical results of operations does not permit a real-time connection between the results and the numbers.

Meanwhile, in a large corporation, data from each work floor are entered into a computer system. These data are then sent to the accounting department and automatically tabulated. Yet, in many cases, work floors do not receive feedback on these results. In many companies, the results might, at most, be sent to executives to show "This is what happened over the month," with work floors receiving no feedback at all. Therefore, work floors in some companies have absolutely no idea of the state of the company.

Even if the income statement and other accounting reports were taken to the work floors with the aim of informing employees of the realities of operations, for most people they would be extremely complex and difficult to understand. There would be little sense of the direct correlation between the statements and their own work.

I devised the hourly efficiency report as a way of getting around this problem. The idea behind the report is to enable all units to gain an understanding of their revenue and expenses in a simple manner, as easily followed as the household accounts kept by the average family.

At the start, amoeba leaders entered only figures of actual results in the tables. Before long, they began to include numerical forecasts at the start of the month. Now, each amoeba produces hourly efficiency reports showing detailed numerical forecasts of its scheduled activities on a monthly basis. They are managing profits while comparing the forecasts with the results of revenue and expenses actually generated through business activity.

In addition, the hourly efficiency report highlights business activity results using added value as a measure. I will explain this in more detail later in this chapter. In the meantime, please note that *added value* here refers to the balance remaining after material costs, depreciation of machinery and facilities, and all other expenses incurred in making the product (*except labor costs*) are deducted from sales revenue. To easily understand how much added value each amoeba is creating, added value per unit hour is expressed as hourly added value per employee, and is arrived at after dividing total added value by total labor hours. This is the indicator called *hourly efficiency.*

Using the hourly efficiency indicator and other means, each amoeba sets monthly and annual targets in managing its results. In other words, with accurate awareness of the results of their own activities in the shape of monthly added value, amoebas can quickly pinpoint problem areas and implement remedial actions.

The Difference between the Standard Cost Accounting Method and the Amoeba Management System

Production departments in many manufacturing industries use standard cost accounting as their accounting method. For factory management, cost accounting has an important role in production cost management, inventory assessment, evaluation of production department results, and other areas.

Many of the electronics manufacturers with strong ties to Kyocera also use standard cost accounting. Some time ago, I heard that when they purchase electronic components from Kyocera and other suppliers for assembling TV sets, for instance, accounting personnel trained in costing calculate the cost of the final product as follows.

They begin by calculating costs incurred in the previous fiscal period. For the current period, they might give instructions to the effect of "These were the costs for the last period. For this period, set cost targets at 10 percent lower than the last period." Receiving such instructions, Production sets the cost targets 10 percent lower than the previous period and endeavors to work within that range. If Production succeeds in making products within that cost range, they have fulfilled their responsibility. However, there is absolutely no awareness of responsibility for the creation of profits.

Then, when the product is complete, the Sales Department receives the product from the Production Department at standard cost. Adding a margin to the cost price to set a selling price and selling the product are the Sales Department's responsibilities. All of this is subject to the abilities of the Sales Department; however, within that department, some might fail to consider the profit of the company as a whole and carelessly set prices low, saying, "Market competition is tough and we have no choice but to sell with just a small margin over cost." Deduct Sales Department operating expenses from the selling price, and suddenly the product has created a loss. Moreover, in many cases actual prices are set not by directors responsible for sales departments, but by sales personnel who base their decisions on the results of pricing surveys in the key electronic goods retail centers of the Tokyo Metropolitan Area such as Akihabara. By this method, sales personnel with minimal experience in sales have the power to decide the course of company operations!

I remember being astounded when I heard how these major electronics manufacturers are conducting their affairs. Some of Japan's leading manufacturers, staffed with highly talented personnel, are actually entrusting a handful of sales personnel with price setting—a role that has an enormous impact on the company's operations.

Even now, using standard cost accounting, many manufacturers are entrusting price setting, and thereby the course of company operations and profit management, to a few people in Sales. I cannot help thinking about how much the abilities of the majority of employees are wasted, in an enterprise with thousands or tens of thousands of outstanding personnel, by a management system that entrusts so much to so few. Although it's a seemingly systematic mechanism, in fact the abilities of company employees often are not being effectively maximized.

In contrast, under Amoeba Management the market price becomes the starting point. Market prices are related directly to amoebas through Internal Sales and Purchases. Production activity is then based on these internal sale and purchase prices. Additionally, because production amoebas are independent profit centers, they take responsibility for lowering costs to enable them to create profit. That is, production amoebas do not make products according to standard costs handed down from above. All amoebas try to apply inventiveness to reduce costs based on market prices, and are burning with their mission to accrue as much profit as possible.

In a manufacturing enterprise, most employees work in production. Therefore, great differences in the profit awareness of employees emerge between ordinary companies where production departments are only concerned about making products at the given cost price, and companies using Amoeba Management.

Production departments using Amoeba Management do not simply follow costs as with standard cost accounting. Their principal objective is what it should be for a manufacturer: applying inventiveness to create products with added value. Thus, Amoeba Management may be regarded as an innovative management system that challenges the traditional thinking of conventional management accounting.

The Amoeba Form Becomes Visible

When I was president of Kyocera, I always brought hourly efficiency reports whenever I went on a business trip, and used every spare moment to study them. In effect, this gave me direct awareness of the faces of the leaders and subordinates in specific units, and the state of the work being undertaken in every corner of the factory.

I was constantly visiting work floors, which gave me a good understanding of the items being produced, materials and production processes, facilities, production technology, leaders in charge of amoebas, work floor atmosphere, and so on. Therefore, simply perusing the figures in an hourly efficiency report allowed me to visualize the conditions of an amoeba's activity, the realities of the operation, and current problems as if they were projected onto a screen.

Some amoeba units talked to me about their wonderful results, and some screamed desperately for help: "Why is the electricity bill for this amoeba so high? Why are travel expenses so high?" Even in the absence of other information, the hourly efficiency report tells everything.

Categorizing expense items in the hourly efficiency report is an important way to determine the shape of the amoeba. The accounts of companies in general show miscellaneous expenses as higher than other, individual expense items. Originally, *miscellaneous expenses* were called that because they were smaller than all other items; large amounts that cannot be disregarded should not be bundled together under this category. I will describe this point in more detail in this chapter. For now, be aware that the hourly efficiency report lists account items in greater detail than ordinary accounts. Expense items in the report therefore give managers a far more detailed awareness of the realities of their operations.

It is important for company management to undertake operations based on objective analyses of operating conditions drawn from detailed hourly efficiency reports. These complement a good knowledge of daily work floor developments. The hourly efficiency report is the crystallization of the struggles and efforts of employees at the work floor. As such, it is a "mirror" giving an accurate reflection of the amoeba's shape.

Amoeba Power Based on the Bonds of Human Minds

Under Amoeba Management, regardless of the size of the organization, each amoeba is expected to increase added value. However, as I mentioned in Chapter 1, with excessive emphasis on contribution to profits, it can be risky to use monetary incentives linked directly with results to raise employee motivation, as happens in many companies.

Kyocera is a company that, from the beginning, has been managed with the concept of heart-to-heart relationships among all employees as its foundation. Kyocera also believes that individuals have been endowed with abilities and talents so that they can be of benefit to humankind and society. Therefore, amoebas with good results do not put on airs or receive large bonuses in return. Instead, amoebas with wonderful performance receive the intangible honor of genuine appreciation and praise from their colleagues.

Moreover, amoebas are not assessed in terms of monetary comparisons of booking, net production, and hourly efficiency. Each amoeba is valued according to how much it has applied inventiveness to expand those figures. Internal competition among amoebas is not the purpose. Rather, the ideal goal of amoebas is independent development of each amoeba's strength, while simultaneously maintaining harmony with related departments. In other words, instead of taking action based on the self-centered notion of looking out solely for one's own interests, Amoeba Management demands that the minds of all employees in amoebas are united for the greater development of the entire company.

CREATIVITY COMES FROM THE HOURLY EFFICIENCY REPORT

Profit Management by Amoebas

This section outlines how amoebas handle profit management with the hourly efficiency report used by Kyocera, and then explains its features. The mechanism will be described in detail later in this chapter.

Figure 4.1 is an hourly efficiency report for a production amoeba within a component division. Gross Production (A: ¥650 million) is obtained by adding Outside Production (B: ¥400 million) and Total Internal Sales (C: ¥250 million). Total Internal Purchases (D: ¥220 million), representing the cost of components and other purchases from amoebas within the company, is subtracted from Gross Production to get Net Production (E: ¥430 million). The last figure is total revenue for this production amoeba.

Added Value (G) is the profit created by the amoeba. This figure is the balance remaining after subtracting the total of all expenses, other than amoeba labor costs—shown as Deductions (F: ¥240 million)—from Net Production (E: ¥430 million). Recorded as Added Value, the difference between revenue and expenses equals ¥190 million and represents the Added Value created by the amoeba. Dividing Added Value by Total Working Hours (H: 35,000 hours) results in Hourly Efficiency for the month (I: ¥5,428.5).

Figure 4.2 is the hourly efficiency report for the sales amoeba of a component division. Total Net Income (C: ¥28 million), or the total revenue of the sales amoeba, is the total of sales commissions on the *custom-order method* and *stock sales business method*. However, this example concerns only the custom-order method and does not include the stock sales business method.

Total Net Bookings (A: ¥360 million), Gross Sales (B: ¥350 million), and Total Net Income (C: ¥28 million) represent income for the sales amoeba. Subtracting Total Cost (D: ¥12 million) from Total Net Income (C: ¥28 million) results in Added Value (E: ¥16 million) for the sales amoeba. Divide Added Value by Total Working Hours (F: 2,000 hours) to obtain Hourly Efficiency for the month (G: ¥8,000).

Calculating Hourly Efficiency makes it possible for each amoeba to see exactly how much Added Value it is creating per hour and to reflect this knowledge in its business activities. The following section describes the features of this hourly efficiency management system and clarifies its main points.

Item			(Unit: Yen, Hour)
Gross production		A	650,000,000
Outside production		B	400,000,000
Total internal sales		C	250,000,000
	Consumer products sales	C1	0
	Consumer products purchases	D1	0
	Raw ceramic and parts sales	C2	60,000,000
	Raw ceramic and parts purchases	D2	30,000,000
	Raw materials and body sales	C3	95,000,000
	Raw materials and body purchases	D3	90,000,000
	Firing sales	C4	32,000,000
	Firing purchases	D4	30,000,000
	Plating sales	C5	0
	Plating purchases	D5	0
	Processing sales	C6	60,000,000
	Processing purchases	D6	60,000,000
	Other sales	C7	2,000,000
	Other purchases	D7	10,000,000
	Internal instruments and services sales	C8	1,000,000
	Internal instruments and services purchases	D8	0
Total internal purchases		D	220,000,000
Net production		E	430,000,000
Total deductions		F	240,000,000
	Raw materials usage	F1	20,000,000
	Metal parts usage	F2	3,000,000
	Cost of sales—purchased goods	F3	3,000,000
	Auxiliary materials usage	F4	2,000,000
	Miscellaneous incomes for auxiliary materials	F5	-200,000
	Internal consumable tools and instruments	F6	1,000,000
	Dies	F7	6,000,000
	General subcontractors	F8	30,000,000
	Cooperative subcontractors	F9	30,000,000
	Supplies	F10	7,000,000
	Consumable tools and instruments	F11	20,000,000
	Repairs and maintenance	F12	9,000,000
	Power and water	F13	10,000,000

FIGURE 4.1
Example of hourly efficiency report: Production amoeba. *Continued*

Item		(Unit: Yen, Hour)
Gas and fuel	F14	6,000,000
Packing materials	F15	2,000,000
Freight out	F16	2,000,000
Temporary labor	F17	5,000,000
Employee expenses	F18	1,000,000
Royalties	F19	0
Maintenance service	F20	10,000
Travel expenses	F21	2,000,000
Office supplies	F22	300,000
Communication expenses	F23	200,000
Public charges and duties	F24	2,000,000
Research expenses	F25	10,000
Professional fees	F26	0
Design charges	F27	10,000
Insurance expenses	F28	300,000
Rent	F29	900,000
Miscellaneous expenses	F30	2,860,000
Miscellaneous income and losses	F31	-200,000
Gains and losses from sale of fixed assets	F32	-1,000,000
Internal interest—capital	F33	5,000,000
Internal interest—inventory	F34	10,000
Depreciation	F35	20,000,000
Internal consumable tools and instruments	F36	5,000,000
Common expenses	F37	-400,000
Indirect department expenses	F38	6,000,000
Internal royalties	F39	200,000
Sales commission and headquarters fees	F40	40,000,000
Added value	G	190,000,000
Total working hours	H	35,000.00
Business hours	H1	30,000.00
Overtime	H2	4,000.00
Common hours	H3	40.00
Indirect departments hours	H4	960.00
Hourly efficiency	I	5,428.5
Production per hour	J	12,285

FIGURE 4.1 ***(Continued)***
Example of hourly efficiency report: Production amoeba.

Items			(Unit: Yen, Hour)
Total net bookings		A	360,000,000
Gross sales		B	350,000,000
Custom orders	Sales	B1	350,000,000
	Sales commission		28,000,000
	Sales income	C1	28,000,000
Stock sales business	Sales	B2	0
	Cost of sales		0
	Sales income	C2	0
Total net income		C	28,000,000
Total costs		D	12,000,000
	Communication expenses	D1	260,000
	Travel expenses	D2	980,000
	Freight out	D3	3,500,000
	Insurance expenses	D4	130,000
	Customs clearance and handling charges	D5	360,000
	Sales commission	D6	360,000
	Sales promotion	D7	0
	Sales rebates	D8	28,000
	Advertising and publicity	D9	130,000
	Entertainment	D10	84,000
	Professional fees	D11	12,000
	Subcontractor and service expenses	D12	20,000
	Office supplies	D13	40,000
	Public charges and duties	D14	75,000
	Rent	D15	560,000
	Depreciation	D16	130,000
	Internal interest—capital	D17	120,000
	Internal interest—inventory	D18	19,000
	Internal interest—accounts receivable	D19	3,000,000
	Cost of sales—purchased goods	D20	0
	Internal consumable tools and instruments	D21	390,000
	Temporary labor	D22	56,000
	Employee expenses	D23	390,000
	Consumable tools and instruments	D24	210,000
	Repairs and maintenance	D25	95,000

FIGURE 4.2
Example of hourly efficiency report: Sales amoeba.

Continued

Items			(Unit: Yen, Hour)
	Gas and fuel	D26	15,000
	Power and water	D27	37,000
	Miscellaneous expenses	D28	110,000
	Miscellaneous incomes	D29	-250,000
	Miscellaneous losses	D30	0
	Gains and losses from sales of fixed assets	D31	0
	Headquarters fees	D32	530,000
	Common expenses	D33	49,000
	Indirect department expenses	D34	560,000
Added value		E	16,000,000
Total working hours		F	2,000.00
	Business hours	F1	1,800.00
	Overtime	F2	100.00
	Common hours	F3	30.00
	Indirect departments hours	F4	70.00
Hourly efficiency			8,000.0
Total net sales per hour			175,000

FIGURE 4.2 ***(Continued)***
Example of hourly efficiency report: Sales amoeba.

Both Sales and Manufacturing Divisions Are Profit Centers

Under Amoeba Management, since both Sales and Production employ self-supporting accounting (effectively becoming "profit centers"), this mechanism induces effort by all members of each amoeba to raise added value—and thus profits—as much as possible.

As I wrote earlier, Production's profit is calculated by taking the value of net production as revenue and subtracting from that figure all deductions except labor costs to get added value. In Sales, added value is the balance remaining after subtracting all expenses except labor costs from total net income. In each case, added value is then divided by total working hours to give hourly efficiency. Therefore, as profit centers, both Production and Sales understand how much added value they are creating and strive to raise the level of their figures.

Subtracting all expenses and deductions other than labor costs from revenue gives added value. Labor costs are not included in deductions and expenses because of their nature. These costs are beyond the control of many amoebas, as they are determined almost entirely by the company's hiring policy and other policies associated with personnel affairs and administration; therefore, amoeba leaders have difficulty exercising control over this expense. Therefore, rather than trying to minimize labor costs that are beyond their control, amoeba leaders focus on time management, which is paramount in controlling productivity.

Hourly efficiency is calculated by dividing added value by total working hours, which yields added value created per hour. Using this figure as a base, Production and Sales strive to raise their levels of hourly profit.

Always Express Targets and Results in Monetary Terms

One feature of the hourly efficiency report is that activity targets and yields are always expressed in monetary terms, not in item volume, so in addition to volume, all in-house slips display monetary value. Therefore, in-house slips display how much they were purchased for, how much they cost to produce, and a quantity base showing how many items were purchased or produced.

Money is a universal measure used by everyone daily, and therefore it provides common ground. To stimulate awareness in employees of the flow of money generated by their work, all slips are designed to show monetary value.

Around the time Kyocera was established, few companies balanced their accounts on a monthly basis. Naturally, with the closing of accounts taking place once every six or twelve months, they had no way of knowing what was happening to profits each month. With a company on the scale of Kyocera, the monthly closing of accounts was novel at that time. Furthermore, the ability to produce monthly profit and loss figures within a week after the monthly closing of accounts was itself a rare feat.

Moreover, Kyocera did not rely on outside accounting firms for accounts closure or processing. The in-house Business Systems Administration Department produced profit tables, so that work floors had a grasp of the day-to-day numerical results on which they based constant improvements.

Using these hourly efficiency reports, I continued to study profit figures carefully. When walking through a factory work floor, for example, if I happened to find raw materials or metal fittings that had dropped onto the floor, I would say to the people responsible, "How much do you think this material costs? Perhaps you think that because it belongs to the company it doesn't matter if it falls onto the floor? If you had paid for it with your own money, you would no doubt have been very distressed to see even one piece land on the floor. That is the frame of mind in which you should conduct your work." If you intend to undertake real work, then it cannot be done with the frame of mind that you were asked to do it, employed to do it, or forced to do it. Whenever I visited work floors, I constantly appealed to employees to modify their outlook and thinking so that they would feel distressed if they saw raw materials that had fallen onto the floor.

Amoeba Management encourages people to value all materials with the same regard given to their personal property and to deal with even the smallest source of wastefulness. The hourly efficiency report records exact amounts down to the smallest monetary unit of one yen, and supports detailed profit management.

Timely Awareness of Unit Profitability

Management is not something to be exercised only after seeing the hourly efficiency report at the end of the month. Monthly profit is the accumulation of detailed figures generated on a daily basis, and we must not neglect efforts to produce profit day by day. Therefore, in terms of hourly profit, vital numerical information relating to bookings, production, sales, expenses, time, and other factors is not held back for summary at month's end. These figures are gathered on a daily basis, and the results are promptly relayed to work floors.

It is the accurate assessment of daily numerical results that enables comparison between the state of progress and the plans. If production falls behind schedule, countermeasures to put production back on track can be considered immediately. The same applies to expenses that exceed planned levels. Steps to bring expenses under strict control can be implemented immediately.

Examining profits on a daily basis for each small amoeba unit makes prompt managerial decisions possible. Daily profit management supports solid plan accomplishment and speedy managerial decisions.

Increasing Awareness of Time Management

Under Amoeba Management, each amoeba endeavors to raise hourly efficiency based on awareness of total working hours, and to improve productivity through the application of creativity.

For example, one section's hourly labor costs average ¥3,600, which breaks down to ¥60 per minute and ¥1 per second. Our work must thus produce added value exceeding labor costs. Workplace members need to have a good understanding of this idea, so it is necessary to raise awareness of time and to create workplaces that are constantly functioning under a certain level of tension.

Of course, reducing total working hours does not mean cutting down the regular working hours stipulated in the conditions of employment. Even if there is no overtime, employees must be active during regular working hours (7 hours 45 minutes). If total net bookings decline and there is only enough work for five hours, the remaining time is naturally included in the calculations. This is when the creative use of time becomes an important factor in a department's management.

If one amoeba has too little work and a neighboring amoeba has too few workers, then excess personnel in the first amoeba can lend a hand in the second. In such a case, the time worked is also transferred, and the total working hours of the amoeba that loaned the employees will fall. In contrast, total working hours will rise for the amoeba that received the help. Overall, time is used effectively.

Kyocera counts time transfers down to thirty-minute increments. Maintaining an accurate awareness of time spent, each amoeba tries to reduce its total working hours as much as possible, thereby improving hourly efficiency.

Modern corporate management values speed above all else. Improving time efficiency is the key to winning in competition. By using the concept of time as an indicator, the hourly efficiency management system within Amoeba Management awakens individual employees to the importance of time, thereby raising productivity. The effect is not limited to raising one's own unit's profit. Ultimately, it contributes to higher productivity for the entire company and greater strength in market competition.

Unifying Operations Management with the Hourly Efficiency Report

Each amoeba's leader and members use the hourly efficiency report as management information to give them a clear view of their master plan, monthly plan, and results. It does not stop there. The figures built up by each amoeba become part of the figures for each higher level organization, namely, for sections, departments, operational divisions, and finally the entire company. Therefore, totaling the hourly profit of each amoeba is the mechanism for understanding the operating performance of the entire company. In addition to numerical results for the company, the master plan and monthly plan data from amoeba hourly efficiency reports are totaled and incorporated into numerical plans for the entire company.

Accordingly, as the whole company shares the common indicator known as hourly efficiency, the format of the hourly efficiency report must be standardized throughout the company to ensure application based on the same criteria and rules. Sales amoebas and production amoebas use different formats because their means of obtaining revenue are different. However, the formats are unified within each department.

This makes it possible to pinpoint problems in even the smallest amoeba and enables top management to steer the right course for operations. Additionally, the results from each amoeba and the company are presented during morning meetings or other occasions, giving all employees a correct understanding of the state of operations in different departments and across the company. The effect is to raise employee awareness of participation in management and to achieve transparent management throughout the company.

FUNDAMENTAL MANAGEMENT RULES BASED ON KYOCERA ACCOUNTING PRINCIPLES

Under Amoeba Management, it is crucial to accurately assess the numerical data generated by amoebas for sales, production, expenses, time, and other factors. To ensure that this is done properly, internal rules for correct and prompt day-to-day processing of accounts must be established and applied. The basis for this concept is the Kyocera accounting principles.

As in all areas of the company, inquiry into the true nature of things based on fundamental management principles is at the heart of Kyocera management and accounting principles. Put another way, decisions are based not on the so-called common sense of accounting, but on the concept "Do what is right as a human being."

A detailed discussion of this subject can be found in my book *Kazuo Inamori no Jitsugaku* (*Kazuo Inamori's Pragmatic Studies: Management and Accounting*), published by Nikkei Shimbun. In order to see how they relate to Amoeba Management, I will give a concise description of the salient points in the rest of this subsection.

The Principle of One-to-One Correspondence

In the course of business activity, products and money are constantly moving about within the company. It is essential to use the hourly efficiency system for an accurate assessment of the flow of these products and money. Therefore, when products and money move, slips must be issued to correspond to each item on a one-to-one basis. At a glance, this may seem obvious; in practice, however, thorough application is not easy.

For example, for companies in general it is common to send goods to customers in advance amid one's day-to-day business activities, and issue corresponding slips several days later. Being very busy, the person in charge may casually decide to send the invoice "within a few days." The matter passes out of his or her mind, and ultimately the money for the goods is never received. This does happen. When money and products are processed separately from invoices, vouchers, and other slips, it becomes very difficult to determine exactly what is happening where, and the consequence is obstruction of business activities.

Allowing this kind of processing is tantamount to allowing fraudulent and malignant practices such as invoice manipulation and off-balance sheet transactions. If it becomes standard practice, then effective management will crumble as morality decays throughout the organization.

The *principle of one-to-one correspondence* aims to prevent such situations through the specification of item-by-item money and product movement, thus enabling transparent administration. When an item moves, without fail some type of slip is issued, checked, and moved at the same time. Anyone can see the one-to-one correspondence of products or money, along with their slips. In other words, slips cannot move unless a corresponding item or money moves at the same time, and vice versa.

Furthermore, the hourly efficiency report must accurately show the operating performance of one month. When an item produced is recorded as a sale in a particular month, the corresponding purchase figures and expenses should be recorded in the same month. If income and expenditures do not correspond, monthly profit will fluctuate substantially from month to month, making it impossible to see the true circumstance of operations.

Strict adherence to the principle of one-to-one correspondence is a prerequisite for accurately assessing operating figures and for preventing fraud and error.

The Principle of Double-Checking Work

All work should be strictly double-checked at all times in all company business. This raises the reliability of company affairs and sustains the company's sound functioning.

This principle emerged from the concept of "management based on the bonds of human minds," which is at the center of my management philosophy. People sometimes seem to be "pushed by the devil" into making mistakes. Let's suppose that the results for this month are especially grim, so in spite of good sense an employee may inadvertently alter a few figures. To protect employees from this kind of human weakness, double checking is established as a management system to prevent dishonesty and error. Two or more people double-check all figures.

The divisional self-supporting accounting system of Amoeba Management induces a strong desire to improve the profitability of one's own unit. For the double-check system to function effectively, organizational systems and rules are vital. Specifically, all business processes, from accepting materials to stocking and shipping goods to collecting accounts receivable, should be checked by two or more people, perhaps in specific job posts, as the work proceeds.

In general, manufacturers set up their purchasing function within production departments. This is probably due to the belief that incorporating this function into production departments aids in fine-tuning specifications and requirements for the quality of materials and goods to be purchased. However, problems such as allegiance to particular business connections can occur when integrated purchasing sections are directly responsible for selecting suppliers and negotiating item prices.

Kyocera has therefore established a Purchasing Department that is independent of its production units. As a result, purchasing units and materials units can point out behaviors that are less than ideal: "Why do you associate only with that one specific vendor? This vendor is cheaper!" Benefits include preventing allegiance to particular vendors. In other words, when multiple units perform a mutual check of the flow of purchased items, this amounts to organizational double checking.

Additionally, deposits and withdrawals of cash; the use of company seals; the management of safes, accounts receivable, and accounts payable; the issue of payment slips; and other matters are always checked by two or more people or job positions. Establishment of this internal checking system ensures the accuracy of operating figures.

The Principle of Perfectionism

Demands for superior product quality today are such that defective goods are altogether unacceptable. In Sales, Production, Research and Development, and all other processes, perfection is demanded.

This applies equally to the achievement of management objectives. Kyocera does not accept the notion that "very close is good enough" in management goals for order bookings, sales, production, hourly efficiency, or other areas. The statement "We didn't achieve 100 percent, but we did reach 99 percent, so let's go with that" is not acceptable. For production and sales targets, full accomplishment is required.

Hourly efficiency reports and financial statements are the basis for managerial decisions. In administrative sections and other areas of management control, even the smallest errors in such figures can lead to poor decisions. Accordingly, nothing less than perfection is required in operating figures.

Achieving perfection is difficult. Even so, the strong desire to achieve it can help one eliminate mistakes and reach targets.

The Principle of Muscular Management

Amoeba Management creates a strong demand for the elimination of wasteful expenses. For that, the company must have a lean constitution. By *lean constitution*, I mean the absence of any excess or anything unnecessary in an organizational structure that is always taut without undue stress. In other words, the company should hold no stock, facilities, or other assets that produce no profit.

In particular, long-term inventory is strictly managed to prevent the generation of nonperforming assets. Items that do not sell are not held in the long term as part of asset capitalization or recorded as alleged assets. In conformity with the realities of the company, unsold items are disposed of quickly to trim assets.

Furthermore, depreciation expenses due to capital investment, personnel expenses, and other fixed costs tend to balloon if left unnoticed. The greatest care is given to ensure these do not increase needlessly. In the case of capital investment, rather than just rushing into a purchase regardless of how wonderful a new facility's performance might be, an organization conducting a thorough investigation determines whether existing facilities can be put to use for the same purposes.

Productivity may rise with the introduction of state-of-the-art facilities; however, this does not necessarily lead to operational efficiency improvements from the perspective of cost effectiveness. Excessive capital investments can weaken an operation's constitution. Fixed costs, once generated, are difficult to reduce. Therefore, extreme caution is needed with capital investments and personnel increases that will raise overhead costs.

The purchase of materials and other items under Amoeba Management functions according to the principle "Buy only what we need when we need it." The idea behind this rule is "Purchase only what is necessary, when it becomes necessary, in the necessary quantities." Purchasing only that which is needed in the immediate future removes waste and raises the level of care given to what is available now. The absence of stock eliminates the expense, space and time needed for inventory control and is economically more efficient. Markets are changing rapidly, and changes in product specifications may mean parts held in inventory become unusable. "Buying only what we need when we need it" eliminates that risk.

The Principle of Profitability Improvement

Enterprises should last long and continue to develop. Pursuit of opportunities for the material and intellectual growth of employees depends upon improving profitability, increasing the amount of cash at the company's disposal, and strengthening its fiscal structure. Improving profitability increases the retained profit of the company and raises the equity ratio while enabling new investment for the future. Improved profitability, when reflected in better company performance, also leads to higher share prices and higher dividends as repayment to shareholders. Therefore, improved profitability can be regarded as a necessary condition for company prosperity.

However, the fundamental operating principle needed to achieve improved profit is actually extremely simple. You only need to devote yourself to "maximizing revenue and minimizing expenses." As stated earlier, under Amoeba Management, the hourly efficiency system is used to apply this principle throughout the company. With the hourly efficiency system, the added value indicator represents the added value created by individual amoebas. To raise added value, we need only "Maximize revenue and minimize expenses" to the greatest degree possible. The level of improvement in amoeba profitability is clearly shown by the hourly efficiency figure, which is obtained by dividing added value by total working hours.

In improving the company's profitability, amoeba leaders must have a strong will and sense of mission to guide the company toward expansion for the sake of greater prosperity for all. This desire must be shared by all members of the amoeba. A united effort by leaders and employees at work floors to raise profits in the course of day-to-day work will ultimately improve the entire company's profits.

The Principle of Cash-Basis Management

Cash-basis management means simplifying operations by focusing on the movement of money. Manufacturers make products, sell them to customers, and obtain money in return. The various expenses incurred along the way are paid out of the money received. Profit is the money remaining after all costs have been paid.

However, in modern accounting, profit for one year is calculated on an "accrual basis." Based on actual sales and expenses generated in that period, revenue from sales is shown even if the money has not been received. Goods and services received during the period are shown as expenses even if they have not been paid for. Therefore, the profit and loss figures shown on the financial statements are not directly connected with the actual movement of money. For the manager who wonders where the profit has been made, the realities are difficult to understand.

Therefore, managerial decisions are best made by returning to the starting point of accounting and focusing on cash—the most important and fundamental element of business activity. For instance, in the hourly efficiency report, purchased supplies and materials are recorded as expenses at the time they are purchased. Money movements stemming from business activities are faithfully reflected in the hourly efficiency reports in the month during which they occur. The outcome is an accounting system that corresponds closely with the movement of cash. Working from cash-basis management, internal accounting systems and hourly profit rules are designed to reduce, as far as possible, any disparity between profit accounting and actual cash on hand.

The Principle of Transparent Management

Accounts should present the unembellished truth about a company, in-house and to the outside world. This means we must use unimpeachable accounting practices when deriving operating figures. They must be transparent and clear to all, from executives to general employees. True figures allow employees to understand the operation's realities, and instill in them a desire to take an active part in management. For executives, because their actions are plainly visible to employees, they are compelled to ensure their actions are fair and highly principled. For publicly traded enterprises, obtaining the trust of investors is extremely important. This requires accurate disclosure of the results of this fair accounting system.

Under Amoeba Management, in our aim for "management by all," we have concentrated on transparent leadership that gives all managers as well as employees a clear understanding of the company's operational realities. At Kyocera, as part of this effort, business results of amoebas and all departments are presented during morning meetings held at the beginning of each month. Furthermore, international management meetings and presentations on management objectives are venues for clarifying circumstances, directions, and issues for the entire Kyocera Group. All of this helps to raise company morale, encourage "management by all," and focus the powers of all employees.

KEY POINTS OF PERFORMANCE CONTROL

One issue in the use of the hourly efficiency system is how accurately and fast it can take numerical results and present a true picture of the business situation for each amoeba. If results are not clearly understood, then the hourly efficiency report cannot reflect the realities of each amoeba. People on the work floor lose the sense that the results are the direct outcome of their work.

Therefore, a unified mechanism and management method are needed for accurately recording numerical results. Before going into detail, I will describe three points that form the basis of management by results.

1. Hourly efficiency reports should accurately reflect activity results according to the functions of amoeba units.
2. Hourly efficiency reports should be fair, impartial, and simple.
3. The flow of business should be recognizable as *results* and *balances*.

Hourly Efficiency Reports Should Accurately Reflect Activity Results According to the Functions of Amoeba Units

Hourly efficiency reports should display revenue, expenses, and time, corresponding to the results of activities undertaken within the functions of the various amoebas. The results must be accurately recorded. A true picture of the operating realities awakens a sense of responsibility for the figures on the part of amoeba leaders and members. This creates a feeling of worthwhile purpose.

Suppose that large expenses not directly related to its activities were imposed on each amoeba by headquarters (HQ). Then the members would be unable to accurately assess the operating situation of their amoebas. Even more importantly, members comprising the organization would fail to develop a sense of connection between their work and the results.

For instance, think of a small-scale amoeba earnestly trying to improve performance that is burdened with large costs imposed by headquarters. In such a case, work floor amoeba members striving on the front lines of business activity could quite conceivably lose any ambition to succeed and improve: "Headquarters is spending too much money and shifting the burden onto us. There's little point in making an effort."

Therefore, clear rules and mechanisms are needed to define all numerical results. They must show which amoeba activities generated the results, how much they generated, and how much they should have generated.

Hourly Efficiency Reports Should Be Fair, Impartial, and Simple

In attaining a clear understanding of the operating situation of each unit, hourly efficiency regulations that unfairly favor a selected amoeba or group of amoebas will not be accepted by employees. Unless the rules for hourly efficiency reports treat all amoebas fairly and impartially from first to last, they cannot possibly function effectively. For instance, clear criteria and rules are needed to specify when and under what conditions products made by different amoebas are to be recorded as *net production*.

It would be very difficult to uphold internal rules that cannot be understood without specialized knowledge and that are complex or in other ways difficult to follow. Therefore, it is important that rules are based on truths and principles, have clear meanings, and are simple. If they are, then all employees will have no difficulty understanding them and consequently taking part in management. In addition, thorough knowledge of and adherence to such rules at the work floor level will raise the accuracy of operating figures.

The Flow of Business Should Be Presented as Results and Balances

In constructing the mechanisms of management by results, simply assessing the figures generated is not enough. They must always be managed in accordance with the flow of business in the form of results and balances. Order bookings, production, sales, and other results always generate a corresponding balance. Numerical data must always be managed in a one-to-one relationship of results and balances.

Orders received from customers are first recorded as *total net bookings.* Until the product is completed based on the order received and recorded as *net production*, the work is managed as part of the *production order backlog.* Next, until the product is shipped by the Sales Department and recorded as net sales, it is maintained under the *sales order backlog.*

Additionally, from the moment the product is recorded as *net production* until it is recorded as net sales, the product is held as *inventory.* After being recorded as net sales, the product is listed under the *accounts receivable balance* until payment is collected. Figure 4.3 illustrates the flow of management by results and balances in the case of custom-order production.

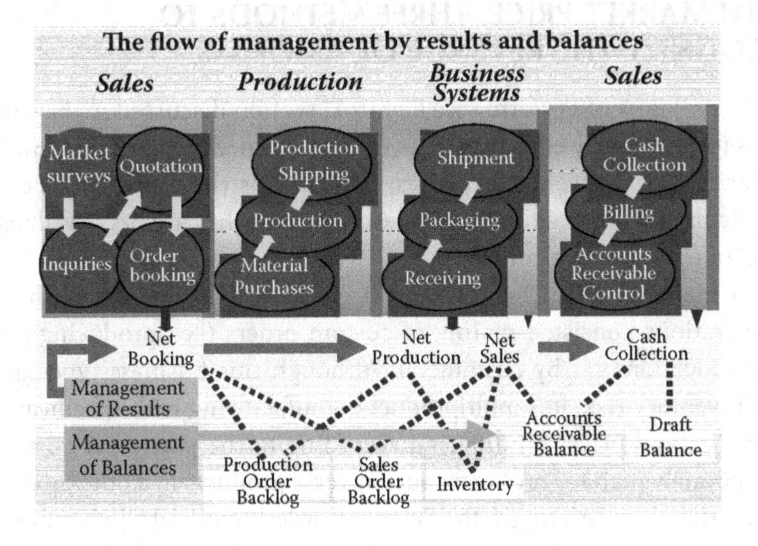

FIGURE 4.3
Flow of custom-order production.

Because the hourly efficiency report for each amoeba shows only results, the backlogs and balances mentioned here are not part of hourly efficiency reporting. Even so, numerical results are always managed together with backlogs and balances, and each amoeba is keenly aware of those numbers. The order backlog, in particular, is a vital management indicator, as sales and production planning are based on it.

The consistent linkage of results and balances ensures there is no incongruity in company operating figures, and maintains one-to-one correspondence at all times. For example, the flow in the custom-order business runs from receipt of the order to production, shipment, and payment collection. For each order, there is a chain of numerical results from *order booking amounts* to *production amounts*, *sales amounts*, and *cash collection amounts*. The corresponding chain of balances created simultaneously runs from *order backlogs* to *inventory* and *accounts receivable amounts*.

Representing the flow of business in terms of one-to-one correspondence allows an understanding of operational realities. At the same time, it clearly sets out the current state of business and forms the basis for correct managerial decisions.

THE CONCEPT OF REVENUE LINKING WITH MARKET PRICE: THREE METHODS TO RECOGNIZE THE REVENUE OF AMOEBAS

The "Key Points of Performance Control" section discussed management issues for the hourly efficiency system. Next, we need to consider how best to clarify actual revenue, expenses, and time as the elements necessary for calculating profit. This section describes how they are perceived, using Kyocera practices as examples.

As I mentioned in Chapter 1, at the time of Kyocera's establishment, its operations consisted mainly of custom orders (i.e., producing goods to specifications set by customers). Although this business type carries little inventory risk, it is multiproduct manufacturing with specifications, deliveries, and prices all differing according to the particular customer. To accurately manage profits for such diverse products in a quickly changing market, we developed the "custom-order" method. This transmits trends in market prices, represented by order values, directly to production amoebas.

Later, Kyocera diversified away from purely custom-order production and developed camera, printer, and other operations. We began holding inventory and selling consumer goods to the general consumer market. With this type of business, sales amoebas forecasted levels of product sales, and took responsibility for holding and selling inventory. In this case, to provide products to the markets in a timely fashion, the Production Department began manufacturing goods on receipt of internal orders from the Sales Department, rather than on direct request from customers.

In contrast to the *custom-order method*, we developed the *stock sales business method* of selling finished products from stock held in inventory. We thus constructed mechanisms, depending on the business type, to gain an accurate picture of the revenue obtained by amoebas. These two systems are complemented by the *internal sales and purchases mechanism*, which facilitates in-house trading between amoebas and provides amoebas with revenue. Thus, Amoeba Management has three methods for defining revenue.

Let us examine each in turn.

The Custom-Order Method

Since the beginning of Kyocera, I have run the company on the premise "The customer decides on the price"—that is, the price is set by the market. Therefore, we do not decide on a selling price after simply totaling all production costs. Instead, we begin by considering movements in market prices, and then try to minimize costs to ensure sufficient profit. In other words, instead of considering [Costs + Profit = Selling Price], our approach has been [Selling Price – Costs = Profit]. In this way, our management style has always been to "Maximize revenues and minimize expenses."

In a market economy of free competition, the market decides the selling price. Based on that selling price, the corporation applies wisdom and effort to reduce costs and produce profit. Although the market price is the base, the market is constantly changing, and market prices can fluctuate from day to day. There is no guarantee that customers are ready to buy this month at the selling price established last month.

To cope with sharp changes in market price, the entire company, not just the Sales Department, must have a good grasp of market trends and work within a system that enables appropriate responses to change. This requires, above all, a profit management mechanism that directly reflects market information throughout the company.

Within manufacturing industries in general, when considering profit management, many companies regard Sales as the profit center and Production as the cost center, with profit managed by Sales. With the Production Department as the cost center, awareness tends to focus on the cost targets preset within the company. In other words, the selling price adopted by the Sales Department has no direct influence on the Production Department. In almost all instances, Production is asked only to cut costs to reach targets. Under such circumstances, a timely response to changes in market conditions is out of the question.

I consider production departments, where goods are actually manufactured, as the source of profits. Production departments should receive market information directly and reflect this information in production activity immediately. For direct synchronization of movements in market prices and revenue accrued by production amoebas within the company, we decided that the sales amount received from the customer should be recorded as the production amount, which corresponds to revenue earned by the Production Department. Meanwhile, as mediator between the Production Department and the customer, the Sales Department would receive a fixed percentage of the value of the sale as commission from the Production Department. For the Sales Department, this commission equates to revenue. (See Figure 4.4.)

In the Production Department, expenses incurred in production activities (sales commissions and manufacturing costs) are subtracted from the

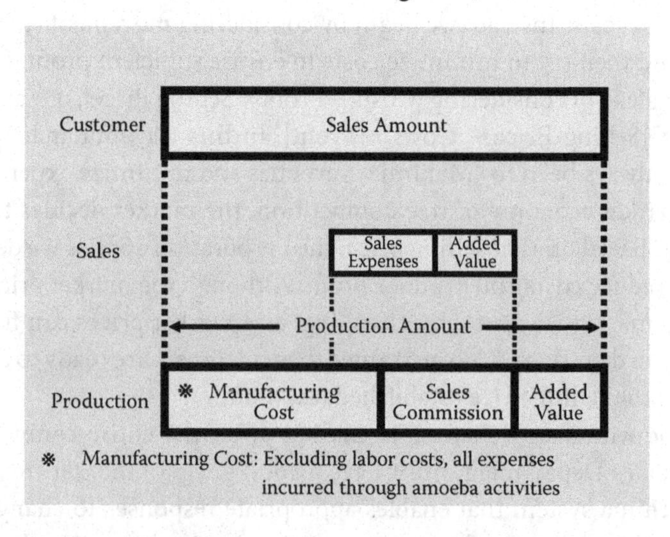

FIGURE 4.4
Income of the custom-order method.

sales amount (production amount accrued over one month) received from the customer. The resulting balance is Added Value – Production. The money remaining after subtracting expenses incurred in sales activities from commission received by the Sales Department over one month is Added Value – Sales.

In defining "Sales = Production Amount by the Production Department," the Production Department is made constantly aware of market prices. Of course, production departments can do more than simply calculate revenue and profit for their own departments. Generally, production departments in other enterprises produce goods within targeted cost ranges, whereas sales departments produce profits. With the Kyocera mechanism, however, sales departments receive a set commission from production departments. From that commission, they deduct expenses incurred during sales activities and endeavor to create a certain amount of profit. The outcome is a thorough awareness that "Production departments are the source of profit." Production departments can thus expand net production according to market movements and reduce expenses to a minimum, which allows them to produce higher profits.

Adopting the Sales Commission Method

At this point, I will explain why I adopted "Commission Received from Production" as the method of earnings for Sales.

When operating an in-house, unit-specific, self-supporting accounting system, one has to consider the method for determining the buying and selling prices between Sales and Production. In the commerce-oriented method, Sales strives to lift its profits as high as possible and buy from Production as cheaply as possible. Meanwhile, Production tries to sell to Sales at the highest price it can. The outcome is intense struggle between the two parties as they try to set a price, with little thought given to the profit of the operation as a whole. The thorough application of self-supporting accounting in amoebas, in this case, would tend to make the relationship between Sales and Production rather awkward.

This is why, for the custom-order method, from the beginning I established the rule that Sales would receive a commission of 10 percent from Production. This method eliminated conflict between Sales and Production over price determination. Of course, I was concerned about the possibility of Sales' morale falling due to the limited commission of only 10 percent. However, Sales is able to raise the absolute value of commission received by increasing sales. Sales thus has the opportunity to lift profit by applying effort. Ultimately, this method has enabled the Sales Departments to produce profit through greater application of effort.

At present, sales commission rates are set according to the type of business and product. In principle, we try not to change rates. If the commission rate were to change for each order, not only would processing become difficult, but also a sense of unfair dealing would emerge due to the lack of uniformity in criteria. If a sense of unfair internal dealings emerges, blame for profit stagnation might be placed on commission rates. The set commission rates should always be regarded as in-house rules, and the pursuit of profit must be undertaken within their bounds.

The Flow of Figures Shows Market Movements

Let us now apply actual figures according to the custom-order method. For example, a product that costs ¥60 is sold for ¥100 in a batch of 10,000 units. The sales amount is ¥1,000,000; the production amount is ¥1,000,000. The Sales Department receives ¥100,000 as its commission of 10 percent. This is the Sales Department's revenue. Meanwhile, the Production Department produces added value of ¥300,000 after deducting production expenses of ¥600,000 and sales commission of ¥100,000. [¥1,000,000 – (¥600,000 + ¥100,000) = ¥300,000].

Suppose now that intensifying market competition forces the product selling price down to ¥90 per unit. The production amount created by the Production Department falls to ¥900,000, and sales commission becomes ¥90,000. The manufacturing cost remains the same at ¥60 per unit, resulting in final Added Value – Production of ¥210,000. Profit suddenly drops by ¥90,000. In short, the moment the selling price falls, the Production Department has a clear understanding of its effect on profits. In response, the Production Department immediately sets out to lower costs in order to recover lost profit.

In contrast, for many companies applying standard cost accounting, the Sales Department buys products from the Production Department using the manufacturing cost as the internal sales price. Therefore, the Sales Department is the only department that is sensitive to falling market prices and has an understanding of profits. Under these circumstances, unless the internal sales price is changed, the Production Department cannot easily respond to circumstances as it is shielded from the direct effects of profit factors. Naturally, simply reducing sales expenses is not a fundamental measure for profit improvement. Ultimately, a delayed response by the Production Department will push down profits further.

Business management today requires speed. Differences in sensitivity to market changes show up directly as differences in corporate strength. Raising the Production Department's awareness of the market equals raising the department's awareness of profit, and continually strengthens its constitution.

Due to the major impact of changes in selling prices on profits, the Production Department must do more than simply reduce costs. Production needs to work increasingly with Sales on the process of price negotiations with customers and on consideration of future trends in orders. The result is the integration of Production and Sales in practice.

The Stock Sales Business Method

For the conventional custom-order method, there was almost no need for a distribution network of retailers and wholesalers, as products were made on receipt of orders and shipped directly to customers. However, Kyocera has diversified into various fields, developing products such as cameras, printers, and recrystallized jewels, which rely on distribution networks for consumer market sales. We thus needed to adopt the so-called stock sales business method of holding and selling goods from inventory (Figure 4.5).

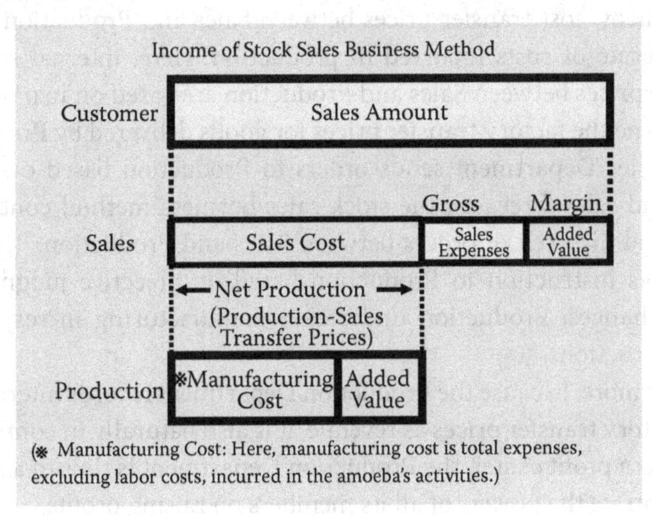

FIGURE 4.5
Income of the stock sales business method.

Under the stock sales business method in Amoeba Management, the Sales Department and Production Department engage in discussions to determine suggested retail prices for goods. They also establish pricing models for each distribution level and decide final selling prices for Kyocera, in addition to internal sales and purchase prices between Sales and Production amoebas.

With inventory selling, revenue for the Sales Department is calculated by subtracting the factory transfer price (the price at which the product is passed on from the Production Department) from the actual sales amount. Also known as *gross margin*, this balance becomes the Sales Department's revenue.

No Transaction at Cost

Manufacturers in general pass goods from production departments to sales departments using cost transfer prices. With this business model, the Production Department predetermines standard costs with reference to past production costs. Based on standard costs, Production then undertakes manufacturing activity as a cost center. Their only concern is with cost management; they do not consider profit. Moreover, as market movements are not directly transmitted to Production, they have difficulty altering cost targets in flexible response to unexpected changes in market price.

In contrast, with the stock sales business method of Amoeba Management, cost transfer prices between Sales and Production are not the aggregate of costs incurred in production. Here, internal sales and purchase prices between Sales and Production are based on market prices and become the factory transfer prices for goods delivered by Production. As the Sales Department sends orders to Production based on market trends and sales forecasts, the stock sales business method controls the receipt and dispatch of orders between Sales and Production. Therefore, Sales gives instruction to Production based on objective judgments of market changes; Production undertakes manufacturing in response to these instructions.

Furthermore, because the Production Department records internal sales at the factory transfer prices as revenue, it is also naturally in control of its profits. As a profit center, the Production Department is thus in a position to concentrate the powers of all its members on raising profits.

With the stock sales business method in Amoeba Management, if the market price falls, then the factory transfer prices to Sales will naturally fall in response. In that case, to reduce negative effects on profits, Production amoebas will independently implement cost reduction measures as a positive step toward improving profits.

Deterioration of market prices is thus directly transmitted throughout the company with both the custom-order and stock sales business methods, and is reflected in each amoeba's profits. Each amoeba has a direct sense of market changes and is able to take prompt action to sustain and improve profits.

Sales Is Responsible for Inventory Management

An important issue with the stock sales business method is how to keep inventory at a minimum to maintain the soundness of company assets. In general, Production can achieve an increase in profit by keeping manufacturing steady. However, simply considering short-term profit and single-mindedly manufacturing goods without considering market movements carry a high risk. By the time you become aware of it, you may have created a mountain of unsellable inventory.

To avoid this situation, under Amoeba Management, Sales alone undertakes uniform management of and responsibility for all stock ordered by Sales, made by Production, and sent to Sales. As part of its responsibility to keep inventory at a minimum, Sales employs precise market analysis. The aim is to make sales and price forecasts as accurate as possible, sending orders to Production only for necessary quantities at appropriate prices. If there are insurmountable discrepancies between sales or price forecasts and reality, and a subsequent write-down of inventory becomes necessary, the Sales Department accepts the fiscal burden of disposal.

Moreover, under the hourly efficiency system, internal interest rates for inventory are set higher than market interest rates and recorded as an expense by Sales. Management of inventory is designed to clarify the responsibilities and liabilities of Sales. Managing inventory and assuming responsibility put Sales in a better position to maintain the company's inventory at a minimum while striving to increase sales.

Minimize Sales Expenses

When selling directly to customers in the custom-order method, the sales commission rate can be set low, as sales expenses are minimal. Sales expenses are adequately covered by the commission, which still leaves a profit.

In the case of selling inventory, stock is sold through distribution channels, such as retail stores. Therefore, inventory risk is high, advertising and publicity are necessary, and sales promotion expenses are required for stores, agencies, and so on. Compared with the custom-order method, selling inventory generates high sales expenses.

Among companies in general, I often hear about departments with high gross margins that can still produce adequate profit compared with other departments even after the deduction of a fair degree of expenses. Much of the higher profit is subsequently wasted on excessive expenses in entertainment and other areas, ultimately lowering the profitability of the entire company.

Under Amoeba Management, whether selling inventory or using the custom-order method, keeping sales-related expenses to an absolute minimum is a fundamental principle. Particularly with sales departments engaging in selling inventory, where sales expenses are inevitably higher compared with custom orders, gross margin rates are also set high. Expenses gradually expand before anyone sits up and takes notice. To prevent this situation, we must constantly identify and cease all wasteful practices and neglect no effort to minimize expenses.

Internal Sales and Purchases

Goods flow through various internal processes before they reach completion and shipment. Under Amoeba Management, in-house transactions between processes are conducted in the same manner as transactions with outside markets. The mechanism for controlling the flow of goods and money between processes is called *internal sales and purchases*.

Generally, even if corporations using division systems buy and sell between divisions at market prices, the goods between processes within Production tend to be passed from one process to the next by cost or working hour (man-hour). The cost incurred in one process is added to the costs accumulated in preceding processes, and passed on. However, this kind of cost-based trading between processes does not occur in Amoeba Management. Even for internal transactions, each amoeba functions as if it is an independent enterprise. Profits are created from in-house transactions, and operations run independently.

Therefore, in the same manner as transactions with outside businesses, materials and works-in-process passed between amoebas are simultaneously recorded as the numerical results of internal sales and internal

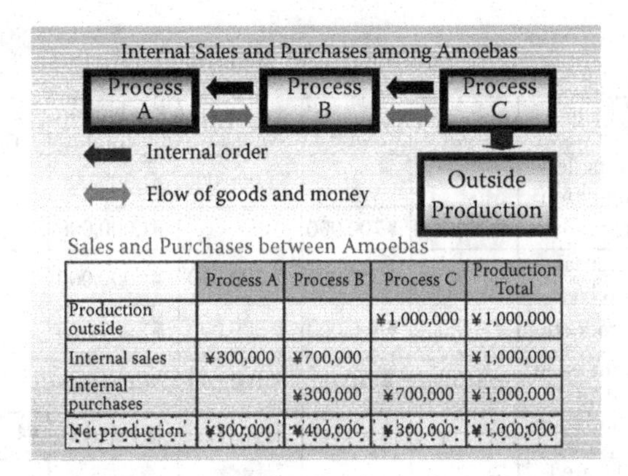

	Process A	Process B	Process C	Production Total
Production outside			¥1,000,000	¥1,000,000
Internal sales	¥300,000	¥700,000		¥1,000,000
Internal purchases		¥300,000	¥700,000	¥1,000,000
Net production	¥300,000	¥400,000	¥300,000	¥1,000,000

FIGURE 4.6
Internal sales and purchases among amoebas.

purchases. Of course, each amoeba considers the good of its own operations and negotiates the best possible buying and selling prices. However, even with in-house transactions it is important for each amoeba to determine prices based on market prices. They must not simply try for prices suited to their respective circumstances. Moreover, all amoebas are evaluated from a market-based perspective on their ability to satisfy internal customers in terms of price, quality, and delivery schedule.

When such market principles permeate every production process, amoebas cannot help but increase their competitive spirit. In addition, amoebas' pursuit of profit establishes the conditions for quality assurance for the next process.

For example, Figure 4.6 illustrates the flow through processes A, B, and C. Process C, which has responsibility for shipment out of the company, receives an order with a value of ¥1,000,000 from the Sales Department. Process C sends Process B an order for works-in-process or necessary materials for the value of ¥700,000. Similarly, Process B presents Process A with an order for materials or works-in-process for the value of ¥300,000.

Process A sends the ordered materials or works-in-process to Process B and records an internal sale of ¥300,000. Net production is now ¥300,000. Process B deducts the internal purchase of Process A from the ¥700,000 received from Process C for internal purchasing. Net production becomes ¥400,000. Process C records outside production for the price of ¥1,000,000 and an internal purchase from Process B for ¥700,000. The subtracted net production thus becomes ¥300,000.

In this way, when goods flow between amoebas, instead of using a cost basis, they are passed across as internal sales and purchases that include

Allocating Sales Commissions

	Process A	Process B	Process C	Production Total	Sales Dept
Outside Production			¥1,000,000	¥1,000,000	
Internal Sale	¥300,000	¥700,000		¥1,000,000	
Internal Purchase		¥300,000	¥700,000	¥1,000,000	
Net Production	¥300,000	¥400,000	¥300,000	¥1,000,000	
Commission Paid	¥30,000	¥70,000	¥100,000	¥200,000	
Commission Received		¥30,000	¥70,000	¥100,000	¥100,000
Allocated Commission	¥30,000	¥40,000	¥30,000	¥100,000	

FIGURE 4.7
Allocating sales commissions.

added value by each process. Following this flow, amoebas run their operations under self-supporting accounting.

Sales Department Commissions

With internal sales and purchases between processes, the production amount equating to revenue is accompanied by payment of an internal commission to the next process, which initiated the order. This means the burden of the final sales commission paid to the Sales Department by the Production Department is divided fairly among all processes.

In fact, when Process A records a sale to Process B, it also records a commission expense of ¥30,000, which is 10 percent of the production amount of ¥300,000. Process B receives the commission from Process A and pays a commission of ¥70,000 to Process C. Accepting the ¥70,000 from Process B, Process C then pays the sales commission of ¥100,000 to the Sales Department (see Figure 4.7).

Using this mechanism, in which the production amount (internal sales and purchase) equals revenue for each amoeba, the sales commission is divided among each amoeba according to fair rules.

Review Profitability of Each Item

Instead of using a cost basis, internal sales and purchases between amoebas are conducted at prices reflecting manufacturing costs and added value at each processing stage. The crucial issue here is the determination of inter-amoeba prices. Price setting is not based on formulas or specific in-house rules. Just as with transactions between companies, prices are determined through negotiations between amoeba leaders. As I mentioned in Chapter 1, based on market prices, amoeba leaders set in-house buying and selling prices with which other amoebas can agree.

At that time, meticulous price determination includes consideration of profit on each item being bought and sold. With regard to orders from customers, internal sales and purchase prices are always set for one-to-one correspondence. It is, therefore, impossible to make questionable decisions on buying and selling between amoebas (e.g., "The price on this item is high, so we'll take a low price for that item to balance it out"). Without one-to-one correspondence for the daily fluctuation of sales prices and costs for each product type, the Production Department as a whole would be unable to exercise control over profit from customer orders.

When setting prices, each amoeba considers its circumstances—market price, productivity, and production yield—and conducts simulations of profit, item by item. For each amoeba leader, that process cultivates talent as a manager.

However, there is always potential for a conflict of interest between amoeba leaders, which can cause internal problems. At such times, people with higher responsibility for both amoebas must constantly be ready to make impartial and correct judgments. If a decision is based on the assertions of only one party, it becomes unfair and may cause both amoebas to lose sight of their obligations toward creating profit. Accordingly, superiors must be ready to listen carefully to the claims of both parties and give guidance from the correct perspective. The superior ultimately has the role of making a fair and appropriate judgment, and of coordinating amoebas' interests for the good of the whole. Therefore, top management, general managers, or others making the final decisions must be of a strong character, and be capable of making correct decisions that will persuade all amoeba leaders involved. In short, they must practice the Kyocera philosophy.

Market Dynamism Is Formed In-House

Buying and selling between amoebas generate more clerical work in the processing of slips, vouchers, and so on. However, the objective of this is to transmit market trends to manufacturing processes, linking them with the outside market through Sales. A selling price that falls when Sales accepts an order from a customer naturally has a major impact on buying and selling between the production processes. Each amoeba is immediately compelled to tackle cost reduction.

Additionally, repeated buying and selling among various amoebas give rise to a market within the company. For example, if there are several amoebas capable of doing the same process, it is possible to deal with the amoeba offering the most lucrative terms.

Furthermore, in some circumstances, it is also possible to send work to outside contractors if cost or quality problems occur with in-house amoebas. Forming an in-house market thus cultivates awareness of inter-amoeba competition, and consequently raises the competitive strength of the company as a whole.

THE CONCEPT OF EXPENSES: CORRECTLY PERCEIVE REALITY AND CONTROL EXPENSES IN DETAIL

As stated throughout this book, "Maximize revenues and minimize expenses" is a crucial element of management. Comprehending the expenses discussed in this section is closely tied with the fundamental management principle to "minimize expenses."

To minimize expenses, top management must take the lead with expense reduction activities. At the same time, all employees on the work floor need to have a strong awareness of the importance of reducing expenses. A prerequisite for this is the existence of a mechanism that gives an accurate understanding of how much is being spent in what areas.

As a tool for work floor profit management, the hourly efficiency report sets out expense categories and narrows down spending to important items. Expense items are detailed on the examples of hourly efficiency reports in Figure 4.8.

RECOGNIZE EXPENSES AT THE TIME OF PURCHASE

A number of rules, explained in this subsection, should be followed when recording expenses under the hourly efficiency system.

Hourly Efficiency Report Items: Production Amoeba		
Item		
Gross production		A
Production outside		B
Total internal sales		C
	Consumer products sales	C1
	Raw ceramic and parts sales/purchases	C2/D2
	Raw materials and body sales/purchases	C3/D3
	Firing sales/purchases	C4/D4
	Plating sales/purchases	C5/D5
	Processing sales/purchases	C6/D6
	Other sales/purchases	C7/D7
	Internal instruments and services	C8
Total Internal purchases		D
Net production		E
Total deductions		F
	Raw materials usage	F1
	Metal parts usage	F2
	Cost of sales—purchased goods	F3
	Auxiliary materials usage	F4
	Miscellaneous incomes for auxiliary materials	F5
	Internal consumable tools and instruments	F6
	Dies	F7
	General subcontractors	F8
	Cooperative subcontractors	F9
	Supplies	F10
	Consumable tools and instruments	F11
	Repairs and maintenance	F12
	Power and water	F13
	Gas and fuel	F14
	Packing materials	F15
	Freight out	F16
	Temporary labor	F17
	Employee expenses	F18
	Royalties	F19
	Maintenance	F20

FIGURE 4.8
Hourly efficiency report items.

Continued

Hourly Efficiency Report Items: Production Amoeba		
Item		
	Travel expenses	F21
	Office supplies	F22
	Communication expenses	F23
	Public charges and duties	F24
	Research expenses	F25
	Professional fees	F26
	Design charges	F27
	Insurance expenses	F28
	Rent	F29
	Miscellaneous expenses	F30
	Miscellaneous income and losses	F31
	Gains and losses from sale of fixed assets	F32
	Internal interest—capital	F33
	Internal interest—inventory	F34
	Depreciation	F35
	Internal consumable tools and instruments	F36
	Common expenses	F37
	Indirect department expenses	F38
	Internal royalties	F39
	Sales commission and headquarters fees	F40
Added value		G
Total working hours		H
	Business hours	H1
	Overtime	H2
	Common hours	H3
	Indirect departments hours	H4
Hourly efficiency		I
Production per hour		J

FIGURE 4.8 (*Continued*)
Hourly efficiency report items.

Continued

Hourly Efficiency Report Items: Sales Amoeba		
Item		
Total net bookings		A
Gross sales		B
Custom orders	Sales	B1
	Sales commission	
	Sales income	C1
Stock sales business	Sales	B2
	Cost of sales	
	Sales income	C2
Total net income		C
Total costs		D
	Communication expenses	D1
	Travel expenses	D2
	Freight out	D3
	Insurance expenses	D4
	Customs clearance and handling charges	D5
	Sales commission	D6
	Sales promotion	D7
	Sales rebates	D8
	Advertising and publicity	D9
	Entertainment	D10
	Professional fees	D11
	Subcontractor and service expenses	D12
	Office supplies	D13
	Public charges and duties	D14
	Rent	D15
	Depreciation	D16
	Internal interest—capital	D17
	Internal interest—inventory	D18
	Internal interest—accounts receivable	D19
	Cost of sales—purchased goods	D20
	Internal consumable tools and instruments	D21
	Temporary labor	D22
	Employee expenses	D23

FIGURE 4.8 (*Continued*)
Hourly efficiency report items.

Continued

Hourly Efficiency Report Items: Sales Amoeba		
Item		
	Consumable tools and instruments	D24
	Repairs and maintenance	D25
	Gas and fuel	D26
	Power and water	D27
	Miscellaneous expenses	D28
	Miscellaneous incomes	D29
	Miscellaneous losses	D30
	Gains and losses from sales of fixed assets	D31
	Headquarters fees	D32
	Common expenses	D33
	Indirect department expenses	D34
Added value		E
Total working hours		F
	Business hours	F1
	Overtime	F2
	Common hours	F3
	Indirect departments hours	F4
Hourly efficiency		G
Total net sales per hour		H

FIGURE 4.8 (***Continued***)
Hourly efficiency report items.

First, for calculating hourly efficiency, all costs associated with the amoeba should be recorded as expenses for the month in which they are generated. For a production amoeba, this includes the purchase of components and materials, electricity charges, equipment interest and depreciation charges, subcontractor costs, repairs, maintenance, and any other expenses relating to production activity. Indirect department expenses and sales commissions are also included in amoeba expenses. However, unlike an income statement, the calculation of hourly added value does not include labor costs as an expense. I will explain this point in more detail further in this chapter.

Second, purchase expenses are handled according to the principle of cash-basis management described in this chapter. All purchase expenses are recorded immediately for the month in which they are incurred. For instance, when an amoeba buys raw materials, all purchase costs are recorded as expenses upon completion of inspection: "Immediately record purchases as expenses." For management of monthly activities and expenses on a cash basis, the expense is not determined according to how much raw material was used in that month, but according to how much was purchased in the month.

However, for devices such as communications equipment, information equipment, cameras, and so on, one model uses a diverse range of components. Component expenses need to be recorded at the time they are actually used; otherwise, monthly profit would suffer wide swings. Moreover, for purchased materials such as expensive precious metals, when there is a difference between purchase lot and quantity of monthly usage, wide monthly profit swings will occur if the entire purchase is recorded in one month.

In such cases, "material expenses on usage basis" may be approved by internal memo, allowing materials and parts to be recorded as expenses in the month in which they are actually used. When using the "material expenses on usage basis" item, it is important to conduct monthly checks to ensure that materials inventory is at an appropriate level and contains no dead stock.

Third, some expenses are not directly connected with amoebas (indirect department expenses, etc.) and cannot be directly controlled by them. These are divided among amoebas in line with mutually agreed upon criteria.

The Beneficiary Principle of Expenses

Under Amoeba Management, units that acquire profits as a result of generating expenses normally have liability for those expenses. This is called the *beneficiary principle.* Expenses directly related to production and sales activities are borne by the relevant unit as a matter of course. Indirect expenses, including the common expenses of indirect departments, are apportioned according to fair criteria. When the beneficiary unit receiving the profit and the amount of expense liability are clearly known, then the expense is recorded by the beneficiary unit, based on the beneficiary principle.

Indirect departments are regarded as cost centers in Amoeba Management. They earn no revenue, and common expenses generated by indirect departments are all transferred to direct departments. In such cases, the expenses are fairly divided according to production amount, shipment amount, number of personnel, floor space, frequency of received benefits, and other features of direct departments. Even then, the principle of one-to-one correspondence is strictly maintained, with the issue of slips or invoices corresponding to the transfer of actual expenses.

To facilitate the movement of their expenses to amoebas, at the start of each month, indirect departments prepare estimates of expenses for that month and report the scheduled expense transfer to each amoeba. Having received notification, amoebas then prepare their own expense transfer schedules.

I hear that many companies are undertaking this kind of expense transfer on a divisional level. Implementing expense transfer on an inter-amoeba level under Amoeba Management raises the accuracy of profit calculation. Extending this practice down to amoeba units with only a few people requires detailed clerical work. However, this practice is essential for enabling people on the work floor to attain accurate knowledge of their operations and to minimize expenses.

Transferring expenses also stirs awareness of spending and helps prevent indirect departments from growing too large. If the actual expenses transferred by an indirect department at the end of the month are much greater than the estimate set at the start of the month, it will naturally affect amoeba profits. Amoebas can then pursue the matter with the indirect department and discover the reason for the increase. Indirect department spending tends to expand. Therefore, by constantly monitoring indirect department spending, Kyocera practices muscular management to cut out wasteful expenses.

If an amoeba leader feels that the transferred common expenses are an excessive burden, thinking, "There will be little improvement even if we cut down on other expenses," then the will to reduce expenses will wither. Under Amoeba Management, the work floor is the center of management, and indirect departments should be very cost conscious, as in small governments.

Handling Labor Costs

The hourly efficiency report was originally a system for calculating how much added value employees create per working hour. I therefore regard people as the source of added value, rather than as costs. Accordingly, instead of handling labor costs as an expense, the hourly efficiency report counts total working hours. Hourly efficiency is then obtained by dividing added value by total working hours.

Of course, labor costs cannot be disregarded in management. Leaders need to be aware of the average labor cost compared with hourly efficiency. If hourly efficiency drops below average labor costs, the amoeba is operating at a loss; if hourly efficiency exceeds average labor costs, it is making a profit. Therefore, each leader must be constantly aware of the break-even point.

However, problems may occur if the labor costs of an amoeba with only a few people are included in the expenses category of the hourly efficiency report. If each amoeba's labor costs are clearly shown, then the profitability of amoebas with members earning high salaries will fall. In contrast, amoebas with members earning lower salaries will show high profitability. In that case, some people are likely to say, "Our profitability is poor because there are members with high salaries." Moreover, there is a risk that with the focus simply on labor costs, amoebas will become unable to function in their primary roles—improving the management and overall operations.

From that angle, the hourly efficiency system targets total working hours for each amoeba, not labor costs, and focuses on profit management from the perspective of hourly added value. Recently, we have been preparing income statements that include labor costs as expense items for amoeba units at production and sales department levels and above. This allows for more detailed calculation of pretax profits, and serves as support material for overall profit management.

Subdivision of Expenses

Earlier, I described the hourly efficiency report as a means for simplifying profit management on work floors. To push down expenses to a minimum, it is necessary to subdivide expense items in the report.

Let us consider the ceramic component production process as an example of why subdivision is necessary. The Raw Material Preparation Unit sells premixed materials to the Forming Unit as an internal sale. The Forming Unit forms the ceramic component and takes it to the kilns in the Sintering Unit. The sintered product is then sent to the next stage.

Suppose there is a goal to reduce electricity charges. However, the *power and water* category includes water and other utility charges combined. It is not clear just how much of the total expense is due to electricity usage. If we set out to reduce electricity charges, we must know what percentage of the expense was generated in which units and processes. Without that knowledge, we cannot know where reductions should be made or clarify the benefits. Therefore, first the power and water category needs to be divided into electricity and water. We then need to ascertain electricity charges for individual units and processes.

To solve this issue, integrating watt meters are installed for raw materials, forming, sintering, and all other processes. Electricity charges are then apportioned according to the quantity actually used in each processing unit, clarifying exactly how much electricity expense is generated by each amoeba. The important point is being able to see, in monetary terms, just how much cost is generated by each unit. If desired, electricity charges can be further reduced with detailed measurement of electricity usage by individual facilities.

Similarly, we may need to cut back on travel expenses for a job with high travel expenses. With all travel expenses bundled together as one item, it is not possible to see what type of travel expenses should receive priority for cutbacks. All travel expense slips are thus gathered and separated into detailed categories for air travel, train travel, taxi fares, accommodation, and so on. It then becomes clear where cutbacks would produce the greatest benefits.

Alternatively, estimates for individual travel expenses could be prepared each month. Leaders could then reduce expenses by directing the use of more cost-efficient travel. Unless expenses are examined at this level of detail, we cannot "minimize expenses." Subdividing expense items in the hourly efficiency report is essential for implementing cost reduction methods that match the realities of operations.

To minimize expenses, the amoeba leader must have sound knowledge of just how the expenses of the amoeba are generated. Without such knowledge, the leader cannot implement concrete measures for profit improvement. Each item in the hourly efficiency report is a vital indicator for understanding the realities of day-to-day management. Leaders must analyze expenses in detail and apply penetrating insight to achieve a reduction in costs.

THE CONCEPT OF TIME FOCUS ON TOTAL WORKING HOURS: BRING A SENSE OF URGENCY AND SPEEDY ACTION INTO THE WORKPLACE

To calculate the hourly efficiency of each amoeba under the hourly efficiency system, we need to know the total working hours of amoeba members. In an hourly efficiency report for production, total working hours is the sum of business hours, overtime hours, internal transfer hours, and indirect department hours over one month for all employees belonging to each amoeba.

If one amoeba assists another, the working time is transferred and recorded as an expense (this type of transfer time is recorded in the report as part of business hours or overtime hours). Total working hours for units common to all sections of a division are also divided among amoebas (internal transfer hours). The same applies to total working hours of indirect departments (indirect department hours), which are apportioned as deemed appropriate. For the information of all members, amoebas receive feedback on total working hours each day, in the shape of time results for the previous day and the total accumulated since the start of the month.

However, labor costs for part-time workers are handled as expenses. They are not time managed. Therefore, labor hours by part-timers are not included in the total working hours on the hourly efficiency report.

Caution is required here. Increasing the number of part-timers and reducing the number of full-time employees raises expenses to a degree, but also reduces total working hours. Outwardly, hourly efficiency appears to have improved. However, the increased use of part-timers for this simple reason cannot be permitted. The use of part-timers requires careful consideration because it can impact future business development and organizational management. Approval by internal memo is needed.

In the hourly efficiency report, since hourly added value is used as an indicator, the concept of time becomes a vital element for determining profitability amid day-to-day business activities. Here it is important to focus on the total working hours of the department, not just on the time spent (time invested) on the actual product manufacturing. This is because time spent on matters other than actual manufacturing activities also has a major impact on profitability. Amoeba leaders and members can thus naturally recognize "the importance of time."

Of course, reducing overtime is not the sole objective of using total working hours as a base. By using the concept of time, each person on the work floor acquires a higher awareness of time in relation to work. This generates a sense of tension and speed, creating a workplace climate in which employees voluntarily strive to improve productivity. Therefore, thorough endeavors to use time more effectively are important means for reducing wasted time by all employees and, consequently, for raising productivity.

5

Fire Up Your Employees

GENERATE PROFIT WITH YOUR OWN WILL: PRACTICING PROFIT MANAGEMENT

The cycle of managing profitability in Amoeba Management is conducted on a monthly basis using the hourly efficiency report. Plans and actual results are prepared each month, and the path from achievement to plan is managed attentively. The basis for the monthly plan is the fiscal year plan, or the *master plan*.

The Fiscal-Year Master Plan

In essence, the master plan represents the will of leaders on how they want to manage their operations during the course of a fiscal year. In effect, it is a summary of objectives for the entire company as well as objectives and goals for individual divisions, prepared after careful and repeated simulations of the likely outcome.

Concrete goals are needed to run a company and lead employees. To clarify management objectives in such areas as sales, net production, added value, and hourly efficiency, it is important to set goals with precise numbers. Moreover, those goals should not simply be figures for the company as a whole. They must be broken down into explicit numbers for each amoeba unit. The reason is that without clear, shared goals, individual employees will tend to aim in whichever directions they perceive as appropriate. The leader is then unable to consolidate the strength of the employees in the direction he or she sets, and the organization as a whole will fail to achieve its objectives.

Although maintaining goals may be desirable, there is little meaning in long-term plans stretching over five or ten years. Amid turbulent changes in the business environment, precise predictions of likely changes in the market are simply not possible. Kyocera has therefore adopted a three-year *rolling-plan system* in preparing for the unclear economic conditions of the future. This is complemented by preparation of the more precise one-year master plan. Kyocera operations are thus based on the rolling plan and master plan. Before the start of each fiscal year, without fail, all amoebas prepare individual master plans based on management objectives and goals set by top management and general managers.

Aligning Our Mental Vectors by Setting Goals

The preparation of master plans begins with careful deliberation by the people with ultimate responsibility for the various divisions or departments, based on company objectives: What role should be taken by the operations for which I am responsible? How much and in what ways should these operations expand? From there, each general manager envisions how operations should grow and advance over the coming year. They must then clearly present these "visions" as concrete objectives and goals, alongside measures for achieving them, to their subordinate amoeba leaders.

Next, working from the objectives and goals of their divisions, amoeba leaders draft master plans for their amoeba units. A master plan must set out more than just goals for sales, production, and hourly efficiency. Drafting the plan from the perspective of managing the amoeba, based on market forecasts and new product release plans for the next fiscal year, the leader in effect prepares a detailed blueprint for month-by-month figures on all related areas, including facilities and personnel. The master plan is not simply a formal plan for increasing output by a certain percentage over the previous quarter. It is based on detailed business planning and strategies, and the plan itself is the result of a series of simulations of possible scenarios.

The master plan figures for each amoeba are totaled for separate divisions. The general manager, as the top person in the division, should then confirm that the figures presented by amoebas concur with his or her aims for the division. If the figures from the amoebas are less than those desired, the manager must convey his or her aspirations to the amoebas and engage in deep discussions about the revision of plans until mutually agreed-upon figures are established. At that time, all amoebas should be fully convinced of the feasibility of their objectives and goals, and therefore be able to summon the ambition needed to attain them.

Derived in this way, the master plan crystallizes general managers' aspirations for their divisions, and the aspirations of the amoeba leaders for their amoebas. Achieving the goals requires strong willpower and a sense of mission to "reach our targets without fail," regardless of what difficulties might stand in the way. I have summarized this as "Maintain an ardent desire that penetrates your subconscious mind." Leaders must have this ardent desire, and it is essential that they share it with their subordinates.

If your mind is focused on management objectives around the clock, then your desire will ultimately filter into your subconscious mind. Sustained desire that is strong enough to permeate the subconscious becomes the driving force behind achieving the master plan. Armed with a strong, burning desire and sense of mission, the amoeba leader repeatedly promotes his or her aspirations to amoeba members. Ultimately, the achievement of the master plan becomes an objective shared by all.

Monthly Profit Management

The monthly cycle of profit management begins at the start of the month. Each amoeba undertakes detailed review of the business environment based on market trends, order status, production plans, and so on, and prepares an activity program.

The monthly plan is a numerical statement of activities the amoeba plans to undertake that month. Therefore, the program contains more than simply calculated sales forecasts and production prospects for the month. It expresses the goals that the amoeba leader wishes to reach and represents a pledge to achieve them.

When preparing a program, it is important to have a good understanding of the results of the previous month and to recognize the problems therein. Based on these reflections, countermeasures for dealing with such problems must be incorporated into the program for the following month. In other words, to achieve the goals in this month's plan, planning must be founded on detailed simulations of what problems can be expected and how to overcome them. Without a clear action plan ready by the time the program has been prepared, achievement of the goals it contains will be difficult.

Monthly Planning Based on the Annual Plan

The goal of profit management through programs and results on a monthly basis is achievement of the master plan. Therefore, when preparing the monthly plan, the main profit categories of sales (production), expenses, added value, hourly efficiency, and others must be verified in relation to the master plan. The aggregate results for the end of the previous month plus estimates for this month must be compared with master plan figures to determine whether progress is proceeding as planned. If the running total falls behind, then amoeba leaders need to ascertain how much the figures should change and implement concrete action to catch up with the master plan.

Overall Verification of Running Totals

Each amoeba undertakes detailed examination of market trends, production plans, expense items, and other areas before placing them in the hourly efficiency report. The forcasted figures are consolidated from the bottom up, from units, to sections, to departments and divisions. Forcasts for the entire company are thus figures totaled from small amoeba units upward. All figures must be backed by sound reasoning.

The program forcasts produced by some amoebas may unavoidably fall behind the master plan. In such a case, the totaled forcasts may fall short of the divisional master plan and thus the all-company master plan. The divisional general manager will then need to carefully analyze the forcasts for all amoebas and reexamine the estimates for the division as a whole. Affirming the conditions in each amoeba, the general manager then guides amoebas that cannot meet their master plan targets through possible ways of making up for the shortfall. Furthermore, if the entire division is in a difficult situation, even amoebas that are meeting their targets are reexamined for ways of raising their forcasts.

The general manager thus does more than simply consolidate the forcasts from each amoeba. As leader of the division, he or she must be ready to work toward achievement of the master plan by one means or another. He or she must be imbued with the challenging spirit needed to strive for even higher goals, and have the power to raise the morale of the division.

Sharing Goals within Amoebas

Having prepared the monthly plan, the amoeba leader must pass details of the program to amoeba members and ensure that the targets are well known and understood.

For amoeba members, making the targets well known and understood equates to making the goals their own. The goals need to be shared by all members to the extent that each person, if asked, is able to recite this month's estimates for orders, sales, production, and hourly efficiency, among other things. Building on that, concrete action plans for achieving the estimates are broken down into targets for individual members.

It is important to note that the achievement of individual targets leads to a sense of accomplishment in fulfilling the goals of the amoeba unit. Therefore, when all members of an amoeba strive to fulfill shared goals, they will also share the enjoyment of accomplishment when those goals are reached. Kyocera has a tradition, dating from the company's founding, of holding *compas* (social gatherings) at such times for members to praise the efforts of others and celebrate the achievement of goals. This stimulates the desire to "work with a strong will" the next month and brings out the motivating spirit of all members to reach even higher targets. Repetition of this activity engenders great energy that can be directed toward achievement of the master plan.

All Members Need an Understanding of Day-to-Day Progress

Numerical results of day-to-day orders, production, sales, expenses, time, and other principal figures are sent as daily reports to each amoeba, reflecting the previous day's results. Amoeba leaders then compare the state of progress with the program and relay the information to members in morning meetings or by other means. In addition, the numerical results for individual units from the previous day are read aloud during all-company morning meetings, informing all employees of the state of orders and production results.

In this manner, confirmation of day-to-day numerical results gives each employee a sense of how the work he or she is currently involved in links to overall results. If results fall behind the estimates, then all members are in a position to examine ways of making up for the shortfall and promptly devise countermeasures. This method concentrates the powers of all members of the amoeba on one goal and leads to the achievement of goals as a group.

Execute Plans with a Powerful Desire to See Their Completion

Leaders must have a strong desire to fulfill the program they have prepared, no matter what it takes. As unit managers, they check daily results, immediately implementing countermeasures for any problems that occur.

It is important for leaders to possess a strong willpower to achieve the plan, whatever it might take, while encouraging their subordinates, and to maintain united efforts by all employees until the end of work on the last workday of the month.

Seen from the perspective of the entire company, one amoeba applying all possible efforts to achieve its goals until the last working moment of the month may make only a very small difference. However, if all amoebas apply the same degree of effort toward the achievement of their goals each month, then the combined outcome will make a considerable difference to the results. Moreover, consistent completion of the programs undeniably heightens the awareness of all employees. This heightened awareness is the power behind expansion of the company's performance.

Together, the strong willpower of the amoeba leader and the accumulation of efforts by all members appear as monthly profits. Therefore, saying, "Last month's profitability was extremely poor; we did not make a profit," reflects the fact that the leader engaged in unprofitable management. The plan for each month should reach 100 percent achievement as the result of the leader's strong willpower and concentration of efforts. A simple excuse for failure is not acceptable.

Additionally, when the month has ended, it is time for the leader to reflect carefully on the activities of the past month: "What steps did we take to complete the program? Were those measures appropriate? Were the measures implemented as planned?" It is important for the leader to distinguish specific management issues from the past month and translate them into management improvements for the next month.

Repeating this process each month leads to improvement in amoeba profitability while simultaneously raising member awareness of participation in management. The steady accumulation of such efforts heightens the leader's managerial mind and helps the leader develop into an outstanding manager. This is a crucial point in the cultivation of leaders within Amoeba Management.

MANAGEMENT PHILOSOPHY THAT SUPPORTS AMOEBA MANAGEMENT

Under Amoeba Management, all amoebas strive to raise their hourly efficiency through daily efforts. They apply three methods: increasing sales (net production), reducing expenses, and using time effectively. Increased sales can be achieved by securing more orders. Expenses can be reduced by cutting waste. Time can be used more effectively by raising labor efficiency.

Although leaders apply these methods to their operations, there are a number of points that are essential in any attempt to improve profits. In this section, I describe some of the most important points.

Pricing Is Management

The proceeds of customer sales are the source of revenue for amoebas. Therefore, with the custom-order method, the level of bookings received from customers greatly affects the profits of production and sales amoebas. The factor determining the level of bookings is the "pricing" of finished products.

When Kyocera was established, the company produced only ceramic components that were used as high-frequency insulating material for weak currents. Uneasy about running an operation on a single product, I approached manufacturers that produced vacuum tubes and cathode-ray tubes, which needed insulating materials, and asked if they had any work for us.

Large manufacturers were already dealing with established ceramic manufacturers. When salespeople from the small, recently established Kyocera visited, they were constantly told, "We will buy from you if your prices are lower." Actually, the response was usually something more like "Prices from another company are 15 percent lower than yours." In response, Sales hastily rewrote the quotes and went back to the customer. With this kind of bargaining, we soon found ourselves compared with competitors simply on pricing.

Sales accepted orders with price reductions of as much as 15 percent. This put great hardship on Production, which was forced to reduce costs even further. I then talked to Sales and said, "Don't you think it is unfair that your easy acceptance of a lower price should force production into such difficulties? If you lower the price enough, then you can receive any number of orders. But this is no reason to praise your sales skills. The mission of Sales is to discover the highest price your customers will pay and still be satisfied with the value they receive. If the price is any lower, then we can receive any number of orders; if it is any higher, then we will lose the order. You have to find that precise level."

If the price is too cheap, then profit cannot be made, no matter how much expenses are reduced. If the price is too high, then goods will remain unsold and turn into a mountain of inventory. Therefore, leaders must thoroughly understand the information gathered by Sales. Working with a sound awareness of trends in the market and of competitors, leaders should price products based on recognition of their value. Pricing is an issue that decides life or death for an operation, and it requires the concentration of the organization's entire nervous system.

Synchronizing Pricing and Cost Reduction

Whether using the custom-order or stock sales method, in the case of products with intense price competition, it sometimes happens that profit cannot be created at the price desired by the customer. Nonetheless, if one is aiming for the long-term expansion of operations, the order is sometimes accepted even though at that moment the selling price undercuts costs and makes profit impossible. At such times, when deciding on a price, we must also consider ways of lowering costs to ensure profit.

These tactics might include purchasing the necessary components at reduced prices or procuring cheaper materials. If that is still not enough, then we need to revise the product design in such a way that it can produce profit. To create profit at the selling price set by the market, it may be necessary to revise even the design and production methods in addition to reducing component costs.

In other words, cost reduction methods need to be considered the moment the leader ascertains that a lower price is necessary. Production must then be notified immediately of the need for drastic cost reductions and the methods to be used for achieving them.

A Leader's Sense of Mission Is Essential in Adapting to Market Changes

Some time back, Systech (later Kyocera's Communication Equipment Division) slipped into an operating crisis and asked for assistance. Systech manufactured calculators and cash registers, among other items. At the time, sales of calculators were rising sharply, particularly in the American market. The Manhattan-based traders who led the U.S. electronics import market made offers to manufacturers such as "If you produce calculators with these functions, we will buy 1 million units." Japanese manufacturers danced to their tunes. They accepted one order after another, expanded factories, increased the number of employees, and set up other systems for increased production. Systech was one enterprise that caught the wave of calculator market expansion and grew rapidly.

However, when the American market became saturated and competition intensified, the situation changed dramatically. American importers began demanding one price cut after another from calculator manufacturers. Suddenly faced with disappearing orders, the Japanese makers who had gone to great lengths setting up production increases were in a panic. Taking this opportunity, American importers demanded even lower prices.

In response to these demands, Systech's president himself negotiated directly with material suppliers in an effort to produce profit. Nevertheless, the company had grown too large, and somewhere along the way the reduction of costs was entrusted to subordinates.

Compelled to comply with incessant demands for lower prices, production leaders took over from the president in demanding lower prices from material suppliers. However, material suppliers had themselves already been lowering their prices for some time and could not simply agree to Systech's requests. In the absence of a strong sense of mission and without a thorough understanding of the need to "obtain materials at this price by whatever means necessary," they were unable to succeed in intense negotiations with suppliers. As a result, profit deteriorated.

However, the number of personnel had risen and the factory had been expanded. People and facilities could not be left idle. While acknowledging the impossibility of the situation, the president was forced to agree to demands from clients for further price reductions. Systech's profit thus deteriorated steadily until operations finally came to a standstill.

What this example tells us is that even if top management decides to lower prices, in time the company may no longer be able to function unless there are in-house leaders who have a will strong enough to draw profit out of operations at those prices. No matter how many management control systems there are, and regardless of how accurate the awareness of operational realities is, in the end it is the people who must respond to falling market prices. In such situations, it is the presence of leaders with a powerful sense of mission to "produce profit even in the face of substantial price declines" that decides the destiny of the company.

Project Our Abilities into the Future

Around the time that Kyocera was established, the custom-order method was the main form of business in this company. Production activity commenced only after orders were received from customers. If the necessary orders were not obtained, then suddenly the work floor was at a standstill with nothing to produce. Therefore, while running operations I was constantly thinking about the size of the order backlog and how much production we could achieve in that particular month.

Eager to increase orders, I visited potential customers to promote our products. In response, we received a request from a large consumer electronics maker for a product that was technically particularly difficult to make. Since many large ceramics makers had been in business long before us, there was little reason to expect that a sales call from a small, unknown, and newly established company would receive any response other than the orders refused by larger companies.

However, the company could not do business by turning down difficult orders. I wanted the order badly enough to say, "We can do it," even though we did not have the technology at that time. I obtained the order. Returning to the company, I told the engineers, "With our technology at the moment, making this product is still too difficult. If we do it another way, then it will be possible. Let's start with the experiments right now." Among the engineers, invariably some would say, "It can't be done!" This sometimes drained the drive and ambition out of everyone else.

In response, I told them, "I agree that it is too difficult at our current level of ability. However, our ability will undoubtedly improve through repeated trial and error between now and the delivery time. Even though we obtained the order by untruthfully saying we can do it, we technically didn't make a false statement if we apply genuine effort to manufacture the product and succeed. Let's pull together between now and the delivery date and make this product."

People who can project their abilities into the future can guide a difficult job to success. With the determination to "turn this dream into reality" and persistent, sincere effort, ability will surely improve and a path will open up.

This concept applies equally to the management of individual amoebas. Leaders are constantly checking the level of the order backlog, which is the foundation of near-term sales, production, and profit, and taking steps to secure future work. Increasing orders requires independent action. Whether the issue is a technological difficulty, or the price of a product being unsuitable for current production methods, leaders can accept the challenge with the understanding that ability is something that will progress in the future. Future-progressive ability makes it possible for all involved to apply earnest effort to the completion of a product, to reduce costs, and ultimately to substantially improve the amoeba's results.

Running Long-Lasting Operations

Under the hourly efficiency system, we lean toward hourly efficiency as an indicator of profitability. However, good hourly efficiency by itself does not mean the operation will continue to do well.

For example, in an hourly efficiency report for a production amoeba, although hourly efficiency may be rising, there may be a sharp decline in the added value ratio (derived by dividing net production into added value). This often happens when the production amoeba consigns most of the production process to subcontractors. The part sent to subcontractors raises subcontractor costs and reduces added value. However, hourly efficiency rises because of a substantial decrease in employee working hours.

In this case, when looking at hourly efficiency alone, the amoeba seems to be doing very well. The reality of operations is rather different. Irrespective of improvement in hourly efficiency, the declining absolute value of the added value created equals a decline in contribution to the company. A small ratio of added value to net production means the business does not have enough strength to create added value and little ability to maintain the current number of employees.

Considered from the angle of safeguarding the long-term employment of employees, leaders must be fully attentive to the necessity of raising the added value ratio, in addition to hourly efficiency. Rather than simply focusing on the hourly efficiency indicator, unit leaders must accurately analyze the realities of their operations from various perspectives, including the added value ratio.

Shortly after starting the company, I too thought that by forming a company with a very small group of knowledgeable people, good profits could be made by applying our collective intelligence and technology to product planning, consigning production to other companies, and then selling those products. This idea is still around. There are companies—ostensibly manufacturers—whose business model entails specializing in technological development, product development and design, sales, and so on, while production itself is consigned to electronics manufacturing service (EMS) production (EMSs are companies contracted to produce electronic equipment).

Although successful for a time, there is no in-house accumulation of technology for product manufacture, which is the heart of the manufacturing industry. Quality problems and other issues will emerge and make long-term success difficult. Ultimately, for the long-term survival of operations and for employee job security, I believe that added value should be created in-house, on the production work floor, where people can devote themselves to product manufacturing.

Under Amoeba Management, it is possible to improve hourly efficiency by increasing subcontracting and reducing the number of employees in an amoeba unit. Yet, in this case, the long-term growth of operations is not possible. Management needs to maintain a long-term viewpoint. In manufacturing, important technology should be kept within the company, with inventiveness leading to higher added value. To keep and accumulate all the fundamental technologies for product manufacturing within the company, Amoeba Management in the manufacturing industry should use subcontracting as little as possible and build in-house integrated production lines that enable the creation of high added value.

Achieving Strong Cooperation between Sales and Production

For each amoeba to raise its own profits, there must be steady communication between Sales and Production, with a maximum exchange of information. Because Sales and Production amoebas are self-supporting accounting units, they are likely to insist upon their particular standpoints, which sometimes results in quarrels. However, as Sales and Production are part of the same company, there is no question of one group succeeding and the other failing. They must stand together.

Because Sales and Production are in the same company—aboard the same ship in a community held together by destiny—they have no alternative to development through cooperation. Unless they work together to provide customers with finished goods and services, total customer satisfaction will not be attained.

Therefore, Sales gives Production accurate market information on moves undertaken by competitors, the kinds of products demanded by customers, how such products are used, their significance to society, and so on. While confirming market trends, orders status, and other factors with Sales, Production assures them that they wish to compare their technological capabilities with competitors' and, thus, provide the market with attractive products at competitive prices. With this type of voluntary and close cooperation, the profits of Sales and Production will rise together, contributing to the development of the entire company. While working together in friendly rivalry, Production and Sales should cooperate in a spirit of consideration for each other.

Always Be Creative in Your Work

Applying oneself diligently to jobs assigned at the workplace is important. However, in Amoeba Management, where independence is prized, this is not enough. As we go about our daily work, we should constantly ask ourselves if the conventional methods are truly sufficient, and whether there is perhaps a better way of doing something. "Today better than yesterday, and tomorrow better than today"—ongoing improvement and enhancement of one's assigned work are at the heart of Amoeba Management.

Kyocera's history of product development illustrates this point. When the company was founded, we started out making U-shaped Kelcima tubes that functioned as insulating material for cathode-ray tubes, exclusively for Matsushita Electronic Components Company. Later, we wanted to sell the same insulator components to other manufacturers, such as Toshiba and Hitachi, and set about cultivating new clients one after another. Moreover, because cathode-ray tubes are a type of vacuum tube, similar components would also be usable as insulating materials in other types of vacuum tubes. We therefore developed new, related products.

After a while, it occurred to us that applications for fine ceramics were probably not limited to the field of electronics, and we subsequently developed a market for industrial machinery parts. In time, with the cultivation of the U.S. market, we began making ceramic transistor headers. Before long, ICs replaced transistors, and, by that time, Kyocera had already developed ceramic IC packages.

Then, around 1980, Kyocera rescued Cybernet Industries, a company producing wireless communications equipment. Kyocera acquired not only the factories but also the engineers. After that, I started a cellular telephone business then called DDI. Our company began developing mobile phone terminals, and with the engineers who had come over from Cybernet, we developed several mobile phones. At present, in addition to mobile terminals, we also manufacture PHS terminals, base stations, and other related equipment. This has grown into an important pillar of Kyocera's operations.

Similarly, our printer business started with a small operation, from which we developed the unique Ecosys line of printers, which use an amorphous silicon drum developed by Kyocera. Today, the fusion of this technology with the Kyocera Mita (now Kyocera Document Solutions) copier technology is supporting the global expansion of operations.

This kind of technological transition was not foreseen at the start. However, because we refused to be satisfied with Kyocera's current state, we applied creativity and inventiveness to cultivate new markets and develop new products in other areas. A resolute acceptance of challenges has built Kyocera into what it is today.

Almost anyone will hesitate to move into a field of endeavor that is outside his or her range of professional skill. However, turning inward and continuing with existing operations until they become fossilized offers no hope of technological progress. If there is a constant, strong desire to create new things, even if specific knowledge is lacking, the borders of technology and business can be breached by consulting specialists in that field, hiring people with specialized knowledge, and other measures.

Not getting stuck in current circumstances and remembering to "always be creative in your work" will enable amoebas to grow and consequently allow the company to develop. This must be the most fundamental action on our agenda.

Set Specific Goals

In management, it is important to set specific targets. In preparation for the master plan, targets represent the manager's aims for sales, net production, and all other items on the hourly efficiency report, on a monthly basis, from the coming April to the following March.

The amoeba leaders themselves put together the planned figures based on their aspirations for their operations, such as "I want to increase the scale of the business, raising sales by 50 percent." After writing down the numerical target for monthly sales, they continue the process, asking questions such as "In selling that much, what will the expenses be? How much profit will we accrue?"

Business Systems Administration totals the results of individual hourly efficiency reports and distributes a summary to each amoeba. However, when it comes to preparing the master plan, the leader of each amoeba must think through the figures he or she wishes to achieve for sales, expenses, hourly efficiency, and other categories before preparing the hourly efficiency report.

These target figures are set according to the leader's wishes for his or her amoeba. The leader must be driven by the strong will to "attain the targets without fail" for them to be of any value. For example, if orders are not high enough to reach a certain level of net production this month, the leader of a production amoeba may decide to accompany sales personnel when they visit customers to obtain orders. A leader must have a will strong enough to implement specific measures to attain the targets for that month.

Additionally, as I wrote earlier, attainment of targets will not be possible if they are not shared by amoeba members. Here it is important for the leader to clarify the specific roles of members and their action goals. In a meeting or an informal celebration, a leader might say, "This year, I want to run the operation in this way. I want sales to grow by this amount. Expenses and time will probably be this much, and I want the hourly efficiency and the profitability ratio to increase to this level. For us to achieve these goals, we need to increase orders by this much. Therefore, I will accompany salespeople on customer visits and work to obtain further orders. I want you to look after things in the factory while I am out."

In this way, the amoeba leader is required to present his or her goals for the year as monthly targets set out in the hourly efficiency report. The leader must think through the concrete actions needed to achieve the targets for sales, expenses, and all other categories, and take resolute action alongside the other members.

Strengthening Each Amoeba

Under Amoeba Management, the company organization is subdivided into many unit operations, called *amoebas,* with management entrusted to amoeba leaders. All amoeba leaders have responsibility for the credible management of their respective amoebas and for fulfillment of the plans they drafted for their amoebas. Moreover, the general manager of a division with many amoebas has responsibility for overseeing the successful completion of all amoebas' plans, while also raising the profitability of the division.

Kyocera does not think that deteriorating profit in one amoeba can be overlooked simply because it is covered by strong profit in another one. For example, when modifying product-manufacturing methods through the introduction of new manufacturing processes or new technologies, the general manager must naturally ensure reduced costs for the processes overall. Before that, though, he must ensure that profit improves in all amoebas at all processing stages. Suppose the introduction of a new technology for one process raises total production profit, even though profitability of that process remains poor. Such a situation cannot be allowed. If a general manager runs operations with this kind of thinking, attitudes will become lax and the profitability of the entire division will suffer. Earnest effort by each amoeba unit to improve operations enables the company to improve as a whole. Amoeba Management stands upon this thinking.

However, ready submission by the leader of any one amoeba unit to the unjust demands of another unit is foolish. Even if an unacceptable demand comes from the general manager, the leader must have the fighting spirit to engage in argument. Management without this fighting spirit is simply inept. While considering the entire company, leaders must also be ready to do everything possible in defense of the organizations entrusted to them.

Sustaining an Awareness of the Entire Company

When practicing Amoeba Management, leaders may try to limit the transfer of top-class personnel for the protection and expansion of their own units. From the perspective of leaders, it may sometimes be difficult to accept a request to transfer an outstanding subordinate to another unit, especially if the employee was trained in the first unit. However, resistance makes it difficult to put the right people in the right places, and hinders the progress of the entire company. Therefore, a leader should maintain the viewpoint of the entire company, putting aside ego and acknowledging that removing organizational limitations placed on outstanding employees will benefit the development of the entire company.

Similarly, when deciding inter-amoeba prices for Internal Sales and Purchases, the first thought should be how it affects the company as a whole. For instance, it sometimes happens that an amoeba leader with a strong nature and a loud voice asserts his or her views and succeeds in setting an unfair price. When I find such an instance, I scold the individual, "You're thinking only of your own affairs. You have no regard for the affairs of the other party. This degree of self-interested thinking disqualifies you as a leader!"

This example and others like it tell us there can be no contradiction between the success of individual amoebas and the prosperity of the whole. One unit alone doing well while the company as a whole deteriorates is meaningless. The amoeba leader must have a strong sense of mission to protect and develop his or her unit. At the same time, the basis of all decisions must be "What is best for the entire company?"

The following episode illustrates how an entire company can be embarrassed by an egotistic leader. It happened in the United States around the time we began developing sales operations there and hiring local people as sales personnel. As described earlier, Kyocera sometimes obtained orders by declaring, "We can do it," when in fact it was beyond our capability at that moment. Occasionally, problems resulted from various issues, such as failure to meet delivery schedules, in which case sales personnel would visit the customers numerous times to apologize and beseech them to wait a little longer. Among the U.S. sales personnel, one person felt this kind of apology would cause a personal loss of face. He calmly told a customer, "Production in our company is a mess."

When I heard that I said to him, "What you said might be true; however, if you say production in Japan cannot be relied upon, then you are effectively destroying confidence in the entire company. Your own affairs are no doubt important. But how does it help your affairs if you drag down the entire company?" This actually happened as I described, and from that point on, I began urging people in Sales, Production, and other areas to always be aware that each of us is a member of the entire company.

As colleagues working in the same company, leaders must have an overview of the entire company, and base their judgments on "Do what is right as a human being." It is assumed that the leader will work to protect and develop his or her amoeba. If the leader does not have the other amoebas' profits in mind and does not give priority to the good of the whole company, then Amoeba Management cannot hope to succeed.

Leaders Should Take Initiative Rather Than Entrust Everything to the Work Floor

Finally, I wish to describe a potential pitfall in Amoeba Management. As leaders in subordinate posts develop, leaders higher up in the organization may be inclined to entrust more work to those raised through the management of amoebas, relegating their own functions to that of flag bearer. Amoeba Management is a system whereby leaders and all employees independently strive to achieve their targets. In the short term, even if some people in the higher ranks withdraw to an extent from "active input," the company might be able to pull through without detriment as long as the lower ranks of an organization are highly capable in their respective fields.

However, the company cannot expect to develop over the long term under such circumstances.

During my time as president, when a problem arose in Sales, Development, or Production, I would place myself at the head of troubleshooting and personally deal with clients and work floor issues as needed. Whenever I had time, I went to the work floor, visiting units that were experiencing difficulties and putting all my effort into finding solutions. Each leader has been entrusted with the management of a unit operation. However, higher leaders should not leave everything to subordinate leaders. They should be well informed of the problems being experienced by individual amoebas and go to the work floor to help find solutions, while giving encouragement to all members.

Moreover, as a manager, I was continually thinking about directions for the company's future expansion, and ways of moving in those directions. I was fully involved in making important decisions affecting the entire company and sought to fulfill high-level duties above those required of a manager.

Seeing managers earnestly striving to accomplish heavy responsibilities for the benefit of others is the stimulus that drives employees to give their utmost to fulfilling their responsibilities for the good of the company. Under Amoeba Management, the higher their positions in the organization, the more that general managers and other leaders with heavy responsibilities must be ready to give twice the effort.

CULTIVATING LEADERS

The Ultimate System to Elevate Managerial Awareness

One factor behind Kyocera's growth has been the outstanding management system called Amoeba Management. However, as I described at the start of this book, Amoeba Management begins to function as it should only when there is a business climate based on the bond of human minds. No matter how rational a management control system might be, it cannot achieve goals unless leaders and members are motivated to use it effectively. The mere existence of an outstanding profit system does nothing to raise profit at the work floor level. Profit improves with the earnest desire of members on the work floor and their independent will to accept difficult challenges.

To fuse an amoeba organization into one life form, the unit leader's thinking and actions are extremely important. First, the leader must have a vision of the outstanding organization he wishes to create from the amoeba. With strong aspirations for the ideal state of the unit, the leader must then be ready to put all his or her energy into achieving that state.

No company ever has enough leaders with such qualifications. Nevertheless, if you select personnel who may be lacking in this area and entrust them with the management of amoeba units, then a sense of responsibility and mission will emerge before long. Leading a unit toward the achievement of goals builds the leader's experience in stimulating the motivation of members and in many other areas. This heightens the leader's ability to understand people's minds and to understand profit management, and it cultivates leadership qualities.

At the same time, in addition to the leaders, organization members also work toward achieving their personal goals, which has the effect of raising managerial awareness in members. In cultivating leaders and heightening the managerial awareness of all employees, Amoeba Management may be regarded as the ultimate training system.

Meetings Are Opportunities for Educational Guidance

In cultivating leaders, appropriate guidance and assessment of each unit's management by senior management (including top management) are important parts of the process. I have been using meetings as places for conducting this kind of practical education. Executive meetings and other management conferences are venues for individual amoeba leaders to present the results of the previous month and plans for the next month based on the hourly efficiency report. Through the content of the presentation and the ensuing discussion, the cultivation of talent takes the form of frank appraisal and guidance on each leader's way of thinking and attitude toward the work.

For instance, after completing a presentation at a meeting, the leader of a production unit was asked by a salesperson when the product would be finished. The production leader replied, "Our target for this product is this date." I then asked him, "Why do you not specify the day by which it will be finished? By voicing your target for completion as this date, you are building yourself a protective wall in case you do not succeed. You cannot be sure of meeting the delivery schedule if you are constantly ready to run away if things go wrong. Unless you have the determination to succeed, no matter what the cost, you cannot expect things to proceed smoothly. The first thing you need to do is change the attitude that causes you to give such a response."

I see language as having the power to express the heart and spirit of people. The remarks of leaders, especially, have a great effect on unit members. Therefore, I have spent much time trying to correct attitudes and ways of thinking by drawing attention to the remarks and comments of leaders.

In meetings, it is important for senior management to obtain a clear understanding of the circumstances of each unit and then discuss how operations should develop and advance. At the same time, senior management needs to guide and cultivate the leaders' attitudes and ways of thinking.

Set High Targets and Live Each Day with Full Effort

If a company sets low targets, then it can achieve only low results. To achieve substantial growth in performance, it is necessary to set high targets.

When Kyocera was still a small company, I told the people I worked with about the grand vision I had for the company: "First, we are going to make Kyocera the No. 1 company in Haramachi [a Kyoto subdistrict]. When we are number one in Haramachi, we will make Kyocera number one in Nishinokyo [district]. Then we will turn the company into number one in Nakagyo Ward, followed by number one in Kyoto. When Kyocera is number one in Kyoto, we will turn the company into number one in Japan, and then the world." At the time, my vision was far removed from reality. Even so, whenever the opportunity arose, I promoted Kyocera as the world's future number one enterprise. As a result, everyone in the company began striving toward achieving high goals.

Meanwhile, I put my energy into achieving the concrete targets of the master plan and monthly plan as short-term goals. "If you live today with all the energy you have available, then tomorrow will come into view. If you apply yourself earnestly to your work this month, then next month will come into view. If you live to the utmost this year, then next year will begin to take shape." A great business can be built only by maintaining high targets and putting one's full effort into achieving them day after day. Aiming for high targets with an unremitting effort to reach them has made Kyocera the global enterprise it is today.

When applying Amoeba Management, it is important that leaders set high targets and give no less than their utmost efforts toward achieving them today, all day, and every day. Leaders should pursue all kinds of possibilities, repeatedly running detailed simulations and then putting all their energy into achievement. As each amoeba becomes able to focus its energy on high targets, the performance of the entire company will undoubtedly improve.

Various problems and issues occur along the path toward achieving the monthly plan, the master plan, and other goals. Leaders must overcome these various difficulties with a will strong enough not to yield and an effort that is not outdone by anyone. Through repeated trials, the leader naturally acquires abilities and ways of thinking that are appropriate for managers.

To work toward high targets and to guide groups correctly, leaders must constantly seek out the correct action, the best decision, and the way things should be. Through repeated trials of this kind, the leader will grow as a human being, while earning the trust and respect of members.

Shared Awareness of the Significance of Decision-Making Criteria

In the Kyocera production departments alone, many diverse amoebas are active. These include high-performance amoebas dealing with flagship products, amoebas that have continued to handle existing products for many years, and amoebas developing new operations. Differences exist in the business environments of the various amoebas. Even so, for each amoeba to expand operations, it must first clarify its business' objectives and significance.

For the leader and all group members to unite as one mind and for the business to move forward, the business must have a legitimate reason. The business must have a higher purpose that gives meaning to its existence and allows it to contribute to the world.

As noted in Chapter 1, the management rationale of this company is "To provide opportunities for the material and intellectual growth of all our employees, and, through our joint efforts, contribute to the advancement of society and humankind." The "growth of all our employees" not only refers to employees but also includes me, as the manager, as well as the management team. At the same time, the company aims to "contribute to the advancement of society and humankind" through technology and business.

I decided on this extremely basic management rationale when the company was founded and I had no experience as a manager. The company was created for the mental and intellectual growth of all who work here, and at the same time for contributing, through business, to the advancement of society and humankind. The company must justify its own existence while striking a chord with its employees, its customers, and all other people involved with it. Therefore, top management must always be ready to clarify to the leaders of the amoeba units "why this business exists"—that is, the underlying meaning and purpose of the operation. Additionally, leaders have a responsibility to ensure that a company's meaning and purpose are reflected in their respective operations. In their own words, they need to ensure that members have a full understanding of the company's significance. Only then will employees be ready to undertake their work as a unified group.

Management is an accumulation of daily decisions, the consequences of which appear as performance results. Leaders, especially, are required to make the right decisions. This means they must constantly strive to follow the universal philosophy "Do what is right as a human being." While acquiring the criteria for making correct decisions, leaders should also work to share these criteria with amoeba members. In meetings, on the work floor, and in any other business scenario, it is essential that leaders are always ready to guide and teach how to make the right decisions and how to solve problems. Ongoing efforts will help to spread the philosophy among members while raising the leader's awareness as a manager.

Section II

Questions and Answers for Management

Q&A Case Studies from Seiwajyuku Seminars

The Amoeba Management I have thus far described will yield its real power when employed in actual management applications. I believe that readers who are in management may be interested in learning how these examples actually apply to specific practical situations.

This part of the book includes questions and answers from the Seiwajyuku, a chain of study groups that I sponsor to help young executives. All Seiwajyuku students are the heads of small to medium-sized businesses. During Seiwajyuku sessions, they present questions on specific management problems that they are struggling with, and I offer my own frank and serious analysis in response.

The following five examples are actual case studies to facilitate the understanding of Amoeba Management. These are actual, specific cases with only minor modifications of the content for inclusion in this book.

QUESTION 1: PROACTIVE INVESTMENT

Our company is a car dealership. Since the Japanese economic bubble collapsed several years ago, we have struggled to rebuild our business. Finally, we are at a point where both our revenues and profit are moving in the right direction. Recently our manufacturer has become very aggressive, pushing a strategy to proactively cope with environmental and safety measures. They plan to double their unit sales in the next four years. They are demanding that each dealer strengthen its sales force and increase sales points.

Our company has three sales offices and fifty employees. Of these fifty, we have fifteen salespeople who sell a total of 450 cars per year. On average, each salesperson sells thirty cars per year, or about 2.5 cars per month.

In the past, our debt has been as high as a third of our annual revenues. We have recently improved our service facilities and made additional financing to meet the manufacturer's request to hold inventory. Since we expect sales to increase, we are planning to build a used-car center, which will be under our direct management and will enable us to offer higher trade-in prices to the customer. As a result of this plan, our debt is expected to double before we can realize our anticipated sales increase.

Although profit is marginally acceptable, we really depend on the manufacturer's incentives. Therefore, we are, to some extent, at the mercy of the manufacturer.

Generally speaking, every Japanese automobile manufacturer is expected to become more selective and more demanding of its dealers in the future. They may set a severe penalty on dealers who do not commit to opening new sales centers or are unable to achieve sales objectives. We believe this will lead to a market shakedown, and we are now facing the crucial moment of whether we will be among the survivors.

We have only two ways to raise our sales: one is to increase our quota on the number of cars each salesperson sells, and the other is to hire more salespeople. There certainly are salespeople out there who can move five or six cars a month. However, these kinds of exceptional salespeople are rare because they have more than just experience and training. We think that hiring and training new people over four years would be safer than hiring more super-salespeople. Those new sales reps should be able to sell 2–3 cars per month fairly quickly.

Establishing a new sales office requires a major capital investment and we have to be careful. We thought of first doubling the number of cars sold by increasing the number of salespeople. Every year, we intend to hire about four new salespeople, which will double our current sales force in four years. That should enable us to sell nine hundred cars per year.

For our company, we consider hiring sales personnel and training them thoroughly to raise customer satisfaction to be absolutely necessary investments that must be done even if it reduces our profit in the short term. Especially in times of economic stagnation, like now, I think we must have the courage to make this type of proactive investment.

On the other hand, there are also many unknowns and risks. We are very concerned about rising competition, sacrificing our profit margins, and putting our survival in jeopardy. I am well aware of Dr. Inamori's advice to "First solidify your footing by improving profitability, then make capital investments." I would like to hear your opinion of our situation.

Comments on Question 1

Seize the best opportunity.
Reduce indirect labor costs and increase profits.

"Be alert and flatten fixed costs" is one of my accounting principles, along with "Always maintain a frugal attitude." Kyocera is a manufacturing company, and our productivity depends upon how advanced our plants and equipment are. However, since our founding, I have been saying, "We should continue to use old plants and equipment." We have extended this frugal attitude toward anything that might increase our fixed costs. In addition, we strictly control our headcount, especially in the indirect labor force (nonsales, nonproduction employees).

According to what you have just explained, you employ fifteen sales representatives, which represents 30 percent of your total workforce of fifty employees. This means that one sales representative must fund the overhead for three employees, including him or herself. And now, you are planning to double your sales in four years by adding four new representatives each year.

As you implement this plan, it will be important to avoid adding any indirect labor. Let us suppose that you only add sales reps who are newly graduated from college, and these new reps have to sell ten cars per year on average in order to add value equal to their salary. You must determine how long the training should be for those employees to enable them to sell ten cars per year. Until they become capable of selling ten cars per year, their contribution to revenue will be lower than their salary. This would be a significant burden. You need to analyze whether your current profits can cover this burden. In addition, you should avoid making such an investment based on the assumption of a future increase in sales. New employees should be hired within your existing financial capacity.

I don't know how long it will take, but you will see quite an improvement in your profit ratio when those new sales representatives each become capable of selling an average of thirty cars per year. Because your indirect labor costs will not have increased, your additional trained sales reps will not have to shoulder the salaries of three employees. If a new sales rep could sell thirty units, the company's profit would equal the salaries of two sales reps. Since fifteen reps currently support a payroll of fifty, it would be much easier if thirty sales reps supported sixty-five, in comparison.

Your operational results indicate that your profit ratio has fallen below a previous level of 5 percent even though sales have increased this year. We must pay extra attention to this. If you consciously accept cost increases that are proportional to sales increases, the costs usually tend to increase more than the marginal increase of sales. Therefore, you should be seriously concerned with the decrease in your profit ratio and set a goal of at least 5 percent profit to sales. It would be a problem if you added new hires at a time when your profit ratio was decreasing. Before hiring anyone, your current expenses must be completely minimized.

Your strong sense of strategy is very good. You should be able to succeed if you continue to have courage and remain very prudent. My suggestion to you is to make a new investment only when you attain at least a 5 percent profit ratio.

You need to watch for opportunities as you expand your business. The business you have just described seems to have a very promising future considering the great importance of safety and the environment. Once your profit ratio rises above 5 percent, if you are sure of this business opportunity, you should go ahead. I have been saying, "Wrestle in the center of the ring," which means that management should always be able to afford taking a good business opportunity without any concerns. You may like to make investments while maintaining a frugal attitude, reducing expenses company-wide. Please make all employees understand the meaning of this investment—a big jump in business—and improve your footing.

Again, indirect labor will also tend to increase if you hire more sales reps. Please never let this happen. You should devote yourself to improving sales while coming up with ideas for improving the repair services structure. Then, your profit ratio will naturally go up along with it.

As your business expands, managers must keep an eye out for details. I would rather not go into specifics here, but store managers and all other employees should calculate their profit ratio for each store, and be thorough in following the principle of "Maximize revenues and minimize expenses." We can't be satisfied with merely "reducing costs." Also, there may be a profit opportunity in the repair service sector. It could turn out that repair services are not just an indirect division supporting sales. In some cases, repair service and sales are divisible, and you can also have a separate used-car center. Each division may have a different structure of profitability, and it will be very crucial to monitor their profitability individually as a manager.

QUESTION 2: ALLYING WITH A LARGER BUSINESS TO INCREASE CAPITAL

I would like to receive your advice concerning capital procurement.

We are a family-owned and family-run hotel business. While the hotel business nationally is stagnant, our company has done well, because our hotels are relatively new and located in good areas with increasing tourism.

But all hotels have the problem of attracting and retaining good employees. It's even difficult to find a good general manager. Because we require irregular work schedules and hard work, we have difficulty even getting qualified customer service people.

However, our hotels have recently been implementing new working arrangements and we have achieved a low level of employee turnover. We also have a successor within the family to take over our business, and our hotels are in good standing.

We are now planning to renovate our hotel, which was built thirty years ago. The layout of the guestrooms should be updated soon in order to meet the needs of families and groups of tourists. The capacity of our guestrooms today is two to three people (for individuals and newlyweds), which is not sufficient for the current trend in tourism. We are planning to build multipurpose facilities, such as conference rooms, banquet rooms, and spas, to accommodate larger groups. If we can double the number of guests we serve, we can triple our revenue. We believe that the effectiveness of this investment will be great enough to generate an acceptable return on the invested capital.

Regarding the most important issue—financing—our financial condition is about average compared to our competitors in the industry. We will not be able to finance this large capital investment from our equity. In fact, we must consider borrowing all of the funds we need for this renovation. Frankly speaking, we cannot expect a bank loan since our debt-to-equity ratio is already high. We have thought about finding a sponsor or establishing a membership, but these alternatives are not very realistic.

The capital investment plan I have just explained is not a simple renovation plan. In commemoration of our thirtieth anniversary, we have adopted a theme to re-establish the old-fashioned style of our business, which is still based on the fixed-rate structure for overnight customers. We will convert our traditional management to a more contemporary and practical cost management system.

Therefore, we would like to implement this plan by teaming up with a large corporation that has recently entered our area. This joint venture would be attractive for this large corporation, too, since they are already planning to build an amusement park in the adjacent area. We will be able to take advantage of their funding resources and credibility now and in the future. Specifically, we will establish a new company through shared capital, and this new company will manage the hotels.

Since we can provide only land and employees, we may transfer part of our managing rights to our joint venture partner. Today's leisure and tourism industry requires not only hearty service at local hotels, but also various promotional activities in major cities, more modern labor management, and effective financial management. We know we cannot stay with family management if we cannot find the appropriate leadership capabilities to expand the business in our own family.

Fortunately, our oldest son is prepared to take over our company, and we have candidates for the general manager and other top positions. The quality of our employees is also high, and many of them would like to have a long career with the company. So, we would transfer only a part of our managing rights to this larger company, if we could contribute to the growth of the area. At the same time, we could keep those good employees in the future and our family members would continue to exercise a role in the management of the business.

May we ask your advice on our capital investment joint-venture plan?

Comments on Question 2

Expanding before improving your profit ratio is risky.
Increase profit to wrestle in the center of the ring.

Frankly speaking, this is a very difficult issue to answer.

You have said, "We may transfer a part of our managing rights to the large corporation if we can use their funds and sophisticated management capabilities." But, later you also explained that you had made a commitment with the oldest son to let him take over management, along with the general manager and others.

I am afraid to sound very severe, but this is not likely to happen the way you want it to. Let us suppose that a large corporation is interested in establishing a joint venture with your company and is so committed, as you have said. In this case, the bank would request a guaranty from the larger corporation, of course. Because this guaranty would be the same as getting a bank loan for the larger corporation, they would probably try to acquire complete management rights from you. The land may be assessed, but it would probably amount to only 10–20 percent of the total loan. You should know that intangible assets, such as your name value and human resources, would not be assessed. In fact, the larger company may try to acquire 80–90 percent of managing rights.

All the employees, including your son, the general manager, and the others, are expecting to be employed for a long time, of course. However, the other company would probably replace your son if he was not highly capable as a manager after a couple of years of service. Even if your son is highly capable, they might replace him once they decide that his management does not meet the requirements of the larger company. The general manager and the others would be in the same position. Therefore, you must be aware that there is no guarantee of keeping the current workforce if you proceed with the type of joint venture plan you have outlined.

I may sound very cold, but the large corporation must act very calmly about these matters since it is investing its money into this project, which will eventually be its responsibility. It is a basic tenet of capitalism to protect your investments.

Of course, your situation is not rare. I feel very agonized in saying this, but with no exceptions, small and medium-sized businesses throughout Japan are experiencing the same problem.

This may be saying it too directly, but seeking a funding partnership with a larger corporation to expand your business based upon the plans you have just outlined seems very risky to me. It may sound harsh, but your operating profit has not grown at all even though your sales have improved a little in the past. Your ordinary income is too low because your interest payments are a heavy burden. On the other hand, your repayment schedule has increased every year. You will need a secured cash flow in order to meet your repayment schedule, even if profit is not significant. Conversely, your debt has increased. Therefore, you must first find a way to improve your present profit ratio and financial condition.

An old proverb states, "God helps those who help themselves." With this in mind, your business has to have the capability of being independent. Otherwise, banks will not loan to you. They will lend an umbrella only to people who can avoid the rain. Unfortunately, your plan to team up with a larger corporation seems to be rooted in your own financial incapability.

The reality is that your profit has not increased at all, even though your debt has, and despite the fact that your sales have increased little by little in a fair volume. It is obvious that your business will face difficulties in the future. You will not be able to resolve the problem by looking for some new way of survival while neglecting to solve the root of the problem. In any case, you need to improve the company's financial condition through your own efforts.

You mentioned that current sales would triple. I think that is unrealistic. Double may be more like it. Various issues, such as staffing and internal management, will naturally rise to make your management more difficult as the business triples in size. I am concerned that your plan is too optimistic. Is it really that easy to achieve threefold revenue—just by rebuilding your hotel?

If you cannot survive with the current thirty-year-old building, you may face reconstruction issues. A thirty-year-old hotel should be fully depreciated by the time you find it becoming decrepit. The plumbing and air-conditioning systems you mention age faster and should be depreciated separately from the building—for about ten years. If you used a fixed rate for depreciation, the building should be almost fully depreciated by now. Your profitability would have improved as a result, and you should have enough retained earnings for future investment. The profit should have been at least 10 percent of revenue and accumulated in your retained earnings.

Soon after I established Kyocera Corporation, I had a chance to hear about "dam-style management" from the late Mr. Konosuke Matsushita. I learned the importance of managing with a strong will to accumulate retained earnings for the company. I put all of my efforts into improving profitability. I did not wait until I was pushed to the edge of the ring to take actions, but acted while there was still distance between myself and the edge. My resolve was strong and I knew there was no possible retreat for me. What you must do now is increase profitability. You need to implement the principle to "Maximize profits and minimize expenses." However, you have to be careful about the possible consequences now, because you may lose customers due to deteriorating service or less elegant food resulting from your necessary cost reductions. You should be the cost leader in your area while upgrading food and service and maintaining your current building. This will require unusual efforts and creative resources.

It may not be a task that anyone can do. If you can do it, you will be an excellent manager. Anyhow, you must completely improve profitability. No matter how attractive the scheme may sound to you right now, I am afraid it will only accelerate the danger to your company unless you increase the base profit first.

Even if there were a corporation that was interested in financing or investing in your plan, you would face an ever-present danger of what to do if the investor should withdraw its invested capital when your plans did not materialize. Overall, it is very dangerous to expand a business by assuming you will have access to another company's capital and credibility.

QUESTION 3: INCREASED DEBT DUE TO EXPANSION

We are in the business of transporting industrial products. Since our founding, we have been able to expand our business by steadily increasing our truck fleet and by setting up a subsidiary to operate warehouse and distribution centers. More recently, our customers have begun to move in droves to a local industrial park, so we set up a processing plant there to support them, since this park does not permit a warehouse without some manufacturing activity. This new investment increased our debt.

The transportation industry is typified by small businesses, and the competition is fierce. To survive, we consider it essential to build up our business and our corporate base. Thus, we have embarked upon an expansion strategy.

However, the basis of a transportation company is steady, hard work. Expanding the scope of the business does not automatically increase our profit. As imported goods replace more and more domestic products, we are also concerned as to whether we can continue to raise the working conditions of our employees. I am worried that perhaps we should resist our expansion strategy and instead reduce our debt, delaying our expansion until after our company has become more firmly established.

Our expansions during the past five years consisted mostly of land for garages, but we have also invested in the processing plant. In five years, we have invested an amount close to one year's revenue.

Our sales during the past few years have been growing smoothly. The profit ratio has also improved with expanding revenues. But interest and depreciation are also increasing. And the processing plant, with low start-up sales, is a long way from becoming profitable. While I understand the urgency of debt reduction, I also believe that it is necessary to make additional investments in our processing plant, which would increase our debt further. However, if we were to continue in this direction, we would never get out of the borrower mentality, even after paying off our current loans. There will always be new investment needs.

Should we continue with this business expansion strategy? Or should I stop the expansion, implement a program to increase organizational efficiency, and build up our corporate reserves? At this critical moment in our decision making, I would appreciate your guidance as to how we should continue.

Comments on Question 3

Understand the figures on your P&L.
Management cannot take place without an understanding of accounting.

This is a very difficult question. First of all, your processing plant and transportation businesses should be separated as individual divisions. You must calculate individual P&L figures for each, including depreciation expenses and labor costs. Then consolidate the two financial statements to see the entire company.

I understand that you have invested mostly in land for garages and your processing plant. Since land does not depreciate, this purchase amount will remain on your balance sheet for good. Therefore, you can make a decision to purchase land from your cash flow, provided that your company has sources of cash for the purchase. If the working capital of your business has been funded, some land can be left unused as long as the company is able to pay the debt and interest, regardless of whether it is funded internally or by a bank loan. However, equipment, industrial products, and construction costs are completely different; depreciation would add to interest expenses.

In other words, if you can obtain enough funding to purchase the land, all you need to bear is the interest. In contrast to this, capital investment in your plant will bring interest expense and depreciation, and continuing profit is absolutely necessary to bear these expenses.

Through your aggressive development efforts, your revenue has steadily grown in the past three years despite fierce competition. Your ordinary profit ratio has grown to 10 percent from 4 percent, while interest and repayments have increased.

Although the business has been expanding, you are concerned about making additional capital investments. You need to judge whether or not the ordinary profit ratio will continue to rise above the present 10 percent level. Your criterion of judgment in making this decision should be whether it will expand not only the scope of your business, but also your profitability.

As the scale of the business expands, all expenses increase—not just the repayment of investment capital. While you strive to hold down costs, you must keep the marginal growth of costs below the marginal growth of sales.

Your current interest expense is very low, but the market trend could change drastically. Interest rates could rise in the near future. Despite those large capital investments, your profit is barely increasing, even though interest rates are presently low. If rates rise, your profit could go down to zero. You might even experience an operating loss. In any case, if your profit starts diminishing, the bank may become reluctant to provide a new loan and may even raise the interest rate of your current loan. Your company could go bankrupt from capital depletion, even if you are making profits.

Therefore, at the end of the month, managers should analyze how a 1 percent increase in interest rates would diminish your ordinary profit. You should also calculate how much rates would have to rise in order to bring your present profit level down to zero. You might want to slow down the expansion after this monthly review.

The principle I am outlining here is that the amount of debt you incur for capital investment should be limited by your repayment capacity—that is, your after-tax profit minus the current repayment amount. In any case, it is very important that you feel a certain uneasiness when you think about expanding your business by increasing debt. Further, if you decide to proceed, it is very important that you have a desire to repay the debt as soon as possible. When I entered management for the first time, one investor supported me with 10 million Japanese yen that he had borrowed from a bank with his house as collateral. I was desperately trying to repay him. One day, he told me, "If you can tolerate the burden of principal repayments and interest, you should borrow more and expand your business, rather than repaying me the principal. That is what makes you an entrepreneur."

Even after that, I still hated the debt mentality and I made repayments as aggressively as I could. He was amazed and he said, "You are indeed a good engineer, but not a good manager." Nonetheless, I wanted to wrestle in the center of the ring without being concerned about debt. Later, I became very fortunate to be able to expand my business without any debt.

So, you are quite right in worrying now. I believe that if you continue to be careful as you are now, your company will continue to grow.

QUESTION 4: SETTING MANAGEMENT GOALS

When I am making our annual or five-year plan, I am often at a loss in trying to set annual sales growth targets and other numerical goals.

For example, in discussing whether our growth should be 20, 25, or 30 percent, I would obviously prefer to set our goal at 30 percent. However, my subordinates have their own ideas, and I need to take into consideration various factors relating to the market and the economy in general. I am finding it quite difficult to pick the correct target.

If I set a goal that is too high, it may be pretty but unrealistic—as we say in Japan, "painted rice cakes in a picture." Yet if I make it too low, I think it will lead to a diminished sense of challenge and vitality in our workplace. Ideally, I want to provide an adequate level of challenge to our employees without making the goal seem unattainable.

When we set a goal, should it be top-down or bottom-up? If it is top-down, employees may consider it just a corporate-mandated "wish." If it is bottom-up, it tends to be only a slight increase over current results.

Please enlighten us on what you consider to be the key in setting the appropriate target.

Comments on Question 4

The plan reflects the top manager's will.

The fact that you are concerned about such a thing is already an admirable trait for a manager. Since setting a goal is the most important factor in management, any executive who takes his or her job seriously will sooner or later encounter this dilemma. You may say it is a dilemma that every successful company is destined to encounter.

Management means doing something with a group of human beings, so we shouldn't talk about management without referring to human minds and hearts, and we shouldn't manage by disregarding the human element.

Therefore, setting a goal for the company is a matter of deciding how to deal with human hearts. It is not easy to do this correctly. If you were to set a target that could not be achieved in spite of your utmost efforts, people would say, "We can just ignore this unrealistic goal." In this scenario, if you were to respond by shifting the goal downward, your employees would think that goals are easily abandoned or revised, and you may find yourself having to further make downward corrections. Once your employees adopt this attitude, you have lost regardless of whether or not you actually lower your goal.

I believe that the role of a manager is to breathe life into a corporation. If we were to think of a company as a human body, a manager is like a brain cell that commands the body to work. When the manager is at the helm actively leading all employees, the company is alive and well. If the leaders start putting priority on their own well-being, they sap the life right out of the company.

Therefore, a management goal must be the willful decision of a person who is seriously concerned about the company without any selfish interest.

In reality, as long as you are worried about how to judge whether the goals are too high or too low, and whether they should be set top-down or bottom-up, you cannot truly excel in your management. If there were one best way to set management goals, anybody could become a good manager.

The problem is not whether the target value is high or low. What is important is that you, the manager, must have a targeted number that you wish to achieve. Then, it becomes a matter of inspiring all employees to act in unison and with vigor toward achieving the target.

If you set a figure that seems too high, your insistence on employees' willpower to achieve it will be met with resignation: "Mr. President, no matter how you say it, what you are demanding is impossible for us to achieve," they will say, or, "Last year's result was negative. You can't ask us to double our output all of a sudden!" Your pep talk will be useless, no matter how strong your own willpower is.

How to appeal to the hearts of human beings is the most important thing in management. This is not just a problem for managers. Whether you are a school teacher, a baseball coach, or any other leader of a group, it is important to understand the psychology of the people involved, and know how you can inspire them to do what you think is the right thing.

To transform the manager's will into the employees' will, there is no other way but to go top-down. Otherwise, you will not be able to set a high goal that everyone will work hard to achieve. It has to be the person on top who says, "Let's double our result next year!" Then, he must explain, "If our company merely continues as it is now, we will eventually lose out. We know of other companies that are growing successfully, such as the XYZ Company." The manager can then motivate the employees by saying, "If they can do it, our company should also be able to do it. We were stagnant until now, but I know we have what it takes to double our sales."

Employees who see the confidence imbued by the person on top may naturally begin to have a greater belief in their own potential. So, when you finally say, "Let's set our goal to be twice as high as last year's result," the atmosphere should encourage all to respond, "Let's do it!"

Every human being, somewhere in his or her heart, has a desire to take on new challenges and break away from the status quo. This heart says there should be more than what there is now. But, everyone also has an opposite thought that it is safer to resist change, to just keep things as they are. If we do something new, it may backfire. And the bigger the group, the less likely they are to want to take on new challenges. If you just leave your employees alone, they will all end up becoming increasingly more conservative.

So, the task of a manager is to draw out that natural tendency that human beings have to challenge something new. You want them to say, "It's worth a try." But for that, the goal has to be daring and worthy of their challenge.

Of course, to set up a truly challenging goal, there has to be a corresponding business opportunity. And, this business opportunity is not something that you can just wait for. You set the goal, decide when and what you want to achieve, and then run many simulations in your head as to how it can be achieved. As you continue to reflect, you start to notice business opportunities. That's where you want to lead your employees. Which of these opportunities should we select? That is the goal you want to set for your company.

In Chinese literature, there is a saying, "Time comes from Heaven, wealth comes from the Earth, and harmony comes through people." Even if you were to gain time from Heaven and wealth from the Earth, things are ultimately decided by the hearts of people. As more people seek to elevate themselves and head toward the corporate goal, even those who were originally unreceptive to the idea start aligning themselves toward the goal. It is a matter of psychology.

QUESTION 5: THE SHORTFALLS OF COST ACCOUNTING

Our company manufactures industrial machinery components and was established only a few years ago. Our market has expanded recently, our production is rapidly expanding, and we now produce a wide variety of products. However, I am concerned that if we continue in this manner, it will eventually become impossible for me to comprehend our true operating results. We want to accelerate the introduction of an accounting control system that will enable us to accurately view and understand management information relating to our own performance.

In the case of manufacturers, I understand that most companies use cost accounting to manage their business. Kyocera, on the other hand, has never used cost accounting. Instead, you developed your own management method based on Amoeba Management.

Would you kindly tell us what you see as the problems of traditional cost accounting?

Comments on Question 5

A manufacturer's profit must come from production.

As you have correctly stated, many manufacturers control finances using cost accounting, especially through what is commonly known as the standard cost accounting method. Kyocera does not do this. Our reason for shying away from the cost accounting approach can be clearly illustrated by an anecdote I once heard about a major electrical equipment manufacturer.

This story came from an executive from that company, which was struggling with a low profit margin. His division manufactured consumer appliances and used a cost accounting system for management control. This system would first project the current retail price for the appliance, including the discounted sales prices, in Tokyo's Akihabara district (known as Electronics Town due to its high concentration of consumer electronics retailers, which sell everything from computers to toasters).

Then, using this final retail price, the system computes margins for retailers and wholesalers, and subtracts Sales, General, and Administrative (SG&A) expenses. Based on this calculation, it produces the "target price" for the manufacturing plant.

If the plant employees work hard to produce the desired products at the target price, they are considered to have met their goal. It does not mean that the plant is profitable, only that it was able to lower its production costs to the target price and that it met the volume requirement. Then sales will purchase the products at this target price, add its own markup, and send them to their distributors for wholesale.

Unfortunately, the market is always full of surprises. When a competing product appears with a slightly different feature, consumers consider all current products obsolete and won't buy them unless they are heavily discounted. When retailers lower their prices, wholesalers naturally ask the manufacturer to lower its price. They say, "Your products are obsolete. Unless you lower your price drastically, we will end up with a huge inventory." This process eventually leads to a price war with deep discounts.

Originally the manufacturer may have planned a 20 percent margin, but this margin decreases rapidly as the retail price drops. They may end up lucky to have a profit of just a few points. In the worst case, as often happens with semiconductor products, a drop in the market creates an avalanche of price erosion, rendering everyone on the supply chain awash in red ink in both manufacturing and distribution.

This executive said that experiencing this cycle over and over created the "conventional wisdom" that no one can make much money in electronic appliances, short of inventing a new "hit" product.

What, then, is the problem? A manufacturer should be making profit in manufacturing. The problem is that they created a system where production no longer has a profit motive, but must merely meet the price and volume targets. Therein lies the problem.

In the current case, a sales executive sets the manufacturing cost target and decides what the price of the product should be. Cost accounting is a way for the manufacturer to determine the selling price, so it uses the cost of production as a way to move the market.

But unfortunately, markets do not move at the will of the manufacturers. Rather, the market ends up deciding the market price regardless of the cost. The suppliers' logic fails, and the manufacturers feel as if the market has betrayed them. The story our executive told us is a clear illustration of this mechanism.

The market economy is becoming a global competition where companies from all over the world compete freely. We must base our thinking on the fact that the market price is something that is determined by the market. A manufacturer must accept the fact that "neither the price nor the cost is fixed," and continuously apply innovative efforts to lower production costs.

Since a manufacturer's profit should be generated at the production level, it is a fundamental mistake to manage production as a cost center, as in the case of this electronic appliance company. In other words, giving a target price to production is tantamount to cutting it off from the reality of the marketplace, and ruins any eagerness to meet the market price. Production will strive to meet the target price, but will have a hard time finding the motivation to lower its costs any further.

Of course, there is a cost accounting method of making profit centers out of production. While this is a far superior system to the previously mentioned method of treating production as purely a cost center, it still has the tendency to isolate production from the marketplace and focus only on attaining the target cost. Therefore, I doubt that it can really meet the true spirit of the premise that "profit must be generated by production."

For example, the standard cost accounting that is widely used as the way to control manufacturing costs would have standard costs assigned to each product. They would be computed for each unit of production or sales. First, a standard cost is given as a target for production to achieve. Production is evaluated by its variance from the target. If the production cost is below the standard cost, we have a positive variance and production is praised for its efforts. If the variance is negative, there is a shortfall. It then analyzes where the discrepancy came from and how it can be improved: raising the material yield, lowering the cost of purchased materials, shortening the process cycle time, and so on. It is seemingly a very sophisticated system. This system produces a nice analytical report that shows how much the actual sales were, what operating profit production would have generated had it met the standard cost exactly, what the variance was, and what the actual operating profit ended up being.

This system appears to consider production as a true profit center. Indeed, many large corporations are using this method. But the method also has shortcomings. First of all, it requires a tremendous amount of labor and expense to compute the standard cost for each product and for each process of production. Then, the task of comparing the actual cost against the standard cost usually falls, not on production, but on an administrative unit, such as a cost accounting department or a cost control department. Since an administrative department does not directly have production responsibilities, it can only set the new standard cost target as a slight improvement over actual past performance.

In other words, this method does not allow production to be a truly autonomous management unit, but places its profit-making task under an administrative unit. The emphasis is on control rather than on management. In the case of a large corporation, this renders the organization highly bureaucratic.

Then, what is one to do? I believe that the answer is to make production a true profit center, as in Amoeba Management. In Amoeba Management, production is directly linked to the market and bears responsibility for pricing its products to meet the ever-changing market price. As it deals with the marketplace, it is enticed to lower its expenses and raise its profitability. Further, production is free to set ambitious targets for sales and profits. The result is a drastically lower cost structure, and all the credit for the effort goes to production, making it the veritable "star" of the manufacturing company.

So, my advice to you is that, even if you are determined to introduce a cost accounting system, you should adapt it to overcome the problems I have mentioned, including the problem of standard cost computation. In any case, you should accept as your premise the concept that "the profit of a manufacturing company should be generated by production." Thus, it is important that you design a system that can be easily understood by production and utilized for their management.

Afterword

For Kyocera Group, the Amoeba Management System is an indispensable management method. All employees are perfectly comfortable using Amoeba Management in day-to-day work. However, until now the thinking and mechanisms underlying the system had not been put into formal writing.

During the years while I was stepping back from the front lines of management, compiling a book that relates the essence of Amoeba Management had been constantly on my mind. Over about five years, I opened space in a busy schedule, gathered Kyocera executives, and conducted Amoeba Management Seminars. A condensed version of the seminar content forms the basis of this book.

By adding the views of Kyocera directors and executives to my lectures, I was able to systematically compile the management thinking and management control method of the Amoeba Management System. Although it may sound like self-approbation, I do believe the management accounting system of Amoeba Management has the potential to break new ground in the accounting field.

Some in Kyocera said that I should not make the Amoeba Management System publicly available. The reasons given were that it is an original management control method I had worked hard to build over many years, and it is the basis of Kyocera's high-profit management. However, due to fervent recommendation by Ms. Minako Hatano of Nikkei Publishing, we decided to publish in the hope that this knowledge might be of service in the development of Japan's economy. Without Ms. Hatano's zeal, this book would not have reached publication. I am also very grateful to Mr. Koichi Ito for his tremendous support in editing the book.

In compiling this book, I am deeply indebted to KCCS Management Consulting President Naoyuki Morita, Vice President Toshiteru Fujii, Director Tatsuro Matsui, Director Takuro Harada, and Seminar Publication Manager Masaaki Hirai for their collaboration. KCCS Management Consulting provides consulting services on Amoeba Management, and has already helped many companies to improve their business performance.

For their support in editing and preparing various materials, I am also most grateful to Kyocera Corporate Officer Yoshihito Ota, Corporate Officer Masakazu Mitsuda, Advisor Hideki Ishida, Education Planning Director Masaki Kozu, Business Systems Administration Department, Planning Section Manager Shoichi Himono, and Shigeyuki Kitani of the Office of Chief Executives, as well as Kyocera Mita (now Kyocera Document Solutions) Corporate Officer Makoto Yoneyama.

Index

A

Accounting principles, Kyocera
  cash-basis management; *see* Cash-basis management
  double-checking work, 73
  muscular management, 74–75
  one-to-one correspondence, 72
  overview, 71
  perfectionism, 74
  profitability improvement, 75–76
  transparency, 76–77; *see also* Transparency
Amoeba leaders, 24–25
  accountability of, 51
  capabilities, 35–36
  character of, 34
  cultivating, 121–122
  development of, 47–48
  impartiality, 31–32, 51–52, 91
  importance of, 121
  initiative of, 119–120
  mission, sense of, 118–119
  profits, goals to improve, 29–30, 31
  willpower of, 108
Amoeba Management
  amoebas; *see* Amoebas
  basis of, 4
  Business Systems Administration Department; *see* Business Systems Administration Department
  conflicts of interest; *see* Conflicts of interest
  decision making, 123–124
  different, creating companies that are, 38–39
  efficiency, 109
  expansion, corporate, 48–49
  flexibility, 26
  goal setting, 116–117, 138–140
  grocery store analogy, 45–47
  hourly efficiency system; *see* Hourly efficiency system
  innovation of, 61
  internal sales and purchases, 13–14
  leadership development, 14–16
  long-term *versus* short-term rewards, 38
  management awareness, cultivation of, 47
  management by all; *see* Management by all
  market-oriented divisional accounting systems, 8–10, 12–13, 49–50, 59–61
  master plan; *see* Master plan
  objectives of, 8
  origins of, 3–4
  philosophy and values of, 34, 54
  profit centers, 67–68
  rationale for, 4–6
  review, continual, 26–27
  sales, increasing, 109
  targets, high, 122–123, 140
  time management; *see* Time management
  transparency; *see* Transparency
  units, division into, 21; *see also* Amoebas
Amoebas. *See also* Amoeba leaders
  accounting units, 23
  conflicts between, 31–32
  cooperation between, 114–115
  corporate objectives, alignment with, 24, 44
  fiscal independence, 23, 44
  functional needs of, 41–42
  goal setting, 104, 106–107
  individual enterprises, 23–24
  labor costs of, 99
  origins of term, 14
  pricing between, 27–29, 91
  profit management by, 63, 89

revenue requirements, 23
self-contained units, 23–24, 44
strengthening of, 117–118
value of organization, expecation to contribute to, 62
verifying forecasts from, 106
visibility of, 61–62
Aoyama, Masaji, 9

B

Beneficiary principle, 97–98
Business Systems Administration Department
assets, management of, 55
feedback of management information, 55
in-house rules, 54–55
infrastructure, role in creating, 53–54
methodology, application of, 55
overview, 53
profit tables, 68
summaries, distribution to each amoeba, 116

C

Case studies
car dealership, proactive investment, 127–128, 129–130
hotel business, capital procurement, 131–132, 132–135
industrial machinery company, shortfalls, 140–144
management goals, 138–140
transport company, debt load, 135–136, 136–137
Cash-basis management, 76
Ceramic production, 12–13
Common-sense accounting, 71
Compas, 107
Competition, free, 81
Competition, global, 142, 143
Conflicts of interest, 29–31
Cost accounting, conventional, 10, 59–60, 142
Creativity, 115–116
Custom-order system, 53, 63, 79, 80
figures, flow of, 84–85
overview, 81–83
sales commission method, use of, 83–84
Cybernet Industries, 115

D

Dishonesty, 32–33

E

Efficiency reports, 7
Effort, value of, 33
Egoism, 51
Electronics manufacturing service (EMS), 113
Ethics, 32–33
Expenses
beneficiary principle of, 97–98
labor, 98–99
minimizing, 57–58, 75, 92, 109, 110
subdivision of, 99–100
time of purchase, recognizing at, 92, 96–97
Extended-family principle, 19

F

Fairness, 33
Family model, 18–19
Frugality, 129

H

Hourly efficiency system
activity results, 77, 78
expense categories, 92
impartiality, 77, 78–79
labor costs, 98–99
management accounting system for, 58–59
operations management, 70–71
overview, 57–58
profit centers, 67–68
profit management, 63, 99
reports, 70–71, 77
targets, 68
working hours, total, 100–101
yields/results, 68

Human minds, management based on, 73

I

Impartiality, 33
In-house rules, 54–55
Internal sales and purchases mechanism, 81
  commissions, sales, 90
  market dynamism, 91–92
  overview, 88–90
  profitability of each item, 90–91
Inventory, 79

J

Japan
  Communist Party, 16–17
  cultural homogeneity, 37
  family model, 18–19
  post-WWII democracy, 16
Justice, 33

K

Kyocera
  accounting principles; *see* Accounting principles, Kyocera
  amoebas, formation of, 44–45; *see also* Amoebas
  disputes, resolution of, 30–31
  founding, 3–4, 18
  growth of, 6, 8, 36, 48–49, 115–116
  hourly efficiency system; *see* Hourly efficiency system
  management rationale, 5–6
  philosophy, 54, 91
  service, consistent, 25
  significance to employees, 6
  values of, 33–34, 91
Kyocera Ceramics, 4, 6, 9

L

Labor costs, 98–99
Labor, indirect, 129
Labor-management relations, 7–8, 16, 17
Leaders, amoeba. *See* Amoeba leaders
Lean constitution, 74

M

Management by all, 35. *See also* Amoeba Management
  objective of Amoeba Management, as, 8
  stimulation of, 6
Management by results, 77
Management consulting, 6
Management decisions, basis for, 10–11
Market dynamism, 91–92
Market economy, 81, 142
Market sensitivity, 14
Market-oriented divisional accounting systems, 8–10, 12–13, 49–50, 59–61. *See also* Accounting principles, Kyocera
Master plan
  day-to-day progress, 107
  goal setting, 104–105
  monthly planning, 105–106, 122
Material preparation, 24
Matsushita Electronic Components Company, 115
Maximize revenues, minimize expenses, 57–58, 75, 92
Mission, sense of, 42–43, 110–111
Miyagi Electric Company, 3, 11
Morale, 37, 106

N

Net production, 78, 79
Nishieda, Ichie, 3–4

O

Old boy networks, 36

P

Payment by results, 37
Perfectionism, 74
Performance control
  business flow, 77, 79–80

hourly efficiency reports reflecting activity, 77, 78
hourly efficiency reports, impartial and fair, 77, 78–79
overview, 77
Performance-based pay, 37
Perseverance, 33
Philanthropy, 33
Pricing, finished products, 109–110
Principles, management, 10–11
Profit management, 63, 99
fiscal-year plan, 103–104; *See also* Master plan
monthly cycle, 105–108
objectives, management, 105
Profitability improvement, 75–76

R

Rough-estimate style of management, 45

S

Sales commission method, 83–84
Sales order backlog, 79
Self-interest, 118
Sell out and buy up method, 29–30
Shofu Industries, 3, 8, 9
Sincerity, 33
Societal common sense, 28
Stock sales business system, 53, 63
adoption of, 85
inventory management, 87
no transactions at cost, 86–87
overview, 86
sales expenses, minimizing, 88–89
Systech, 110

T

Talent, 33
Time management, 69–70, 109
Total net bookings, 79
Transparency, 8, 19–21, 76–77

W

Workers' rights, 17
Working hours, total, 100–101

Printed in the United States
by Baker & Taylor Publisher Services